Human-Led, AI-Enabled

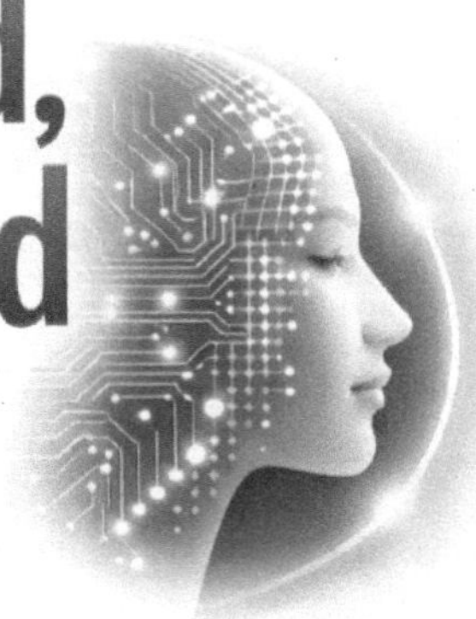

A Practical Guide to Thriving in the AI Era

MOHAMMED BENZAKOUR

This is a work of nonfiction. The information in this book reflects the author's views, experiences, and opinions. Any examples, stories, or references to individuals or events are used for illustrative purposes only. Any resemblance to actual persons, living or dead, or to actual events is purely coincidental. The views expressed do not necessarily represent those of any organization or institution.

ISBN 979-109828771-8

First published 2026

TABLE OF CONTENTS

INTRODUCTION

Since the launch of ChatGPT in 2022, we have entered a new era—one where intelligence is no longer locked inside elite institutions, expensive consultants, or the “right” job title. A person with a phone, an internet connection, and the ability to ask the right question can now access knowledge and capability that used to take years to earn.

I felt the weight of that shift in a recent conversation.

After finishing a talk at an AI conference, I met a man in his mid-fifties who told me his story. He had built a long twenty-year career across education and partnerships management, working with respected organizations. Then he lost his job overnight, and he had been searching for work for two years. He hadn’t used much AI, and he had a limited digital identity and wasn’t active on social platforms. His daughter was twenty years old and also trying to find work. They both described the same feeling: it was like trying to run on a treadmill while everyone else had learned how to sprint—with AI.

I spoke with him about practical steps: how to rebuild visibility, communicate value clearly, and use AI tools to become more productive and relevant. But what struck me most was not the technology. It was the anxiety. The fear of being left behind. The sense that the rules had changed, and nobody had handed them the new playbook.

That story is not isolated. Over the past two years, I've heard the same worries from thousands of people—students, professionals, managers, founders, and parents—trying to figure out what this new age means for their work, their identity, their confidence, and their future.

That is why I wrote this book.

This is not a book about worshiping AI—or about fearing it. It's a book about staying human while the world becomes more automated. It's about using AI as an extension of your ability, without surrendering your judgment, your voice, or your purpose. And it is about not getting left behind when over eight hundred million people, or about 10 percent of the world population, are already using ChatGPT every week.

If you are just starting your AI literacy journey, this book will give you a clear foundation: what AI can do and cannot do, and where humans still matter most. If you are already using AI or experimenting with agents, this book will help you step back and see the bigger picture—how AI reshapes knowledge, expertise, jobs, education, entrepreneurship, art, and mental health. It will also answer the deeper question underneath it all: what gives a human life meaning when productivity is no longer the main scoreboard?

I've used AI extensively. In fact, I am ranked in the top 0.01 percent of users and have built many workflows and agents to test what actually creates results—not just what sounds impressive. I've seen genuine productivity gains, but I've also seen the hidden risks: dependence, confirmation bias, and the slow weakening of the thinking muscle if we outsource too much too fast.

My goal is simple: I want every reader to finish this book able to apply at least one idea immediately—something that strengthens your career, your learning, your self-trust, or your ability to navigate change.

Here is how the book is structured:

- First, we explore AI as an **extension of human capability**—and what it changes about value, expertise, and intelligence.
- Then we move into the **human consequences**: the purpose crisis, the skills shift, and what jobs will demand next.
- Next, we zoom out to the **systems around us**—leadership, education, and entrepreneurship—because your future isn't shaped only by what you personally learn, but also by what institutions reward.
- Finally, we address the **human soul of the era**: art, identity, mental health, and the practical ways to adopt AI without building a cage around your mind.

This book is an invitation to stay in the driver's seat while using AI as a tool to your benefit.

AI can give you speed, scale, and options. But it cannot give you intent. That part remains yours. And in the AI era, your intent—your judgment, values, authenticity, and purpose—will matter more than ever.

ACKNOWLEDGMENTS

It takes days of rain and rivers full of water to fill the sea. The book is a journey that I did not walk alone.

I would like to thank my father and my mother; your values of education, wisdom, and growth have shaped my mind and my thoughts. A heartfelt thank you to my brother, sisters, nieces, and nephews. Your continuous support helped me focus on the vision of this book.

To my professional tribe, I extend my deepest gratitude to my mentors, coaches, peers, and friends who challenged my thinking and who always brought bigger and richer perspectives. Some of our talks sparked my curiosity, opened my heart, and flooded my mind with ideas.

Finally, I want to thank God for giving me the clarity, patience, and inspiration to bring this work to life. I hope that my efforts will serve humanity for the better and for the good.

CHAPTER 1: THE EXTENSIONS OF HUMANS

Despite what many fear or believe, Generative AI is simply an extension of humankind—with all the positives and negatives that come along with it. For centuries, civilization has advanced by creating tools that expand human capability. When the industrial revolution came in the nineteenth century, machinery offered the possibility to work and produce more with less physical effort. Machinery became an extension of our muscles. We ran the machines, repetitive manual work shifted away from human hands, and we were able to scale faster and work more efficiently. That shift changed lifestyles—from handcrafted survival to industrial wealth—and it changed what society rewarded. Value was no longer measured by what you could produce by hand, but by how you could engineer the process that could produce at scale.

Later, in the 1950s, technological progress took over. With the advent of computers, we were able to automate and digitize work. Value shifted again. It became about speed, systems, and how fast we could customize products and services. The digital revolution expanded our capacity to scale knowledge and reach new frontiers.

Through all this change, machines did not replace the human mind. They simply replaced repetition. They created new roles, new industries, and new ways to contribute.

You can see that shift in a profession like accounting. What once took days of paper, organizing, and manual analysis moved into software systems. Over time, less energy could be spent on organizing invoices and more could be spent on interpreting results, identifying risks, and helping businesses grow. The tool didn't remove the need for humans—it moved human value higher, toward judgment and strategy.

Those who embraced the new tools grew with the era. Those who resisted often fell behind.

Generative AI accelerates this pattern. When it became widely available in 2022, adoption spread rapidly across the world. With a new extension, we don't just do the same work faster—we begin solving bigger problems from a different perspective. But this only helps us if we understand what should be delegated to the extension and what must remain in human hands.

If we look ten years ahead, value in society will not be defined by expertise and knowledge alone, but by our ability to navigate multidimensional concepts that combine AI with a specific field. The advantage will belong to people who can use AI for scale while keeping human judgment, direction, and responsibility at the center. For example, I write these words with pen and paper—low-tech tools by today's standards—but I use AI to research. AI is the extension, but I provide the "hand," the intent, which must remain human.

So, can Generative AI—the extension of humans—replace humans? No. Humans created this intelligence. Humans will improve it. Humans will decide what to give it and what not to give it. The more clearly we understand how our potential can be elevated with Generative AI, the more empowered we become to do things better and faster—without giving away the part that makes our work and human perspective meaningful.

Real estate is a good example of how this may look. In a world where Generative AI can run calculations using real-time data, compare scenarios, and build detailed valuation plans, some might assume real estate agents will be replaced. But clients will decide what to hand off to the AI extension and what still requires a human touch. Analytics and finance tasks such as calculating ROI (return on investment) and yield can be delegated to AI because it can do them faster and often more accurately. But subjective criteria—lifestyle priorities, local nuance, long-term context, and the “what’s really happening here” knowledge a good real estate agent brings to the table—still belongs to the people who spend their days on the ground.

Trust is the only remaining scarcity.

AI is an extension of human capability—nothing more, nothing less. It can provide the muscle: speed, scale, and analysis. But the hand has to remain human: intent, judgment, and responsibility. That alignment is what turns AI into an advantage.

Summary

The Core Shift

- **The old way**: Treating new technology as a threat that replaces human contribution.
- **The new way**: Treating Generative AI as an extension of human capability—where the tool provides speed, scale, and analysis, and humans provide intent, judgment, and responsibility.

Key Takeaways

- **The evolution of value**: In the Industrial Era, value came from manual labor. In the Digital Era, it came from speed and systems. In the AI Age, it comes from knowing how to combine AI with domain expertise and human judgment.
- **What tools replace**: Technology consistently removes repetition and lowers the cost of routine work. It does not remove the need for human oversight.

- **The "hand" principle**: AI can support research, drafting, and analysis, but it cannot define priorities, interpret nuance, or own outcomes. Humans still have to be the hand that holds the pen, to decide what matters and what actions follow. AI is the pen, but you are the author.
- **The new advantage**: As more technical tasks become automated, the differentiator moves toward interpretation, context, and decision quality—especially in fields where local nuance and human priorities matter (as in real estate).

The Reader's Audit

If you are reading this to transform your mindset, take a moment to answer these questions:

- What repeatable tasks in my work can be delegated to AI (research, summarizing, scenario comparisons, calculations, first drafts)?

- Where does my work require human-only leadership (intent, judgment, responsibility, context, nuance, priorities)?
- If AI speeds up my analysis, what will I do with the time it returns to improve decision quality and outcomes?

The Next Step

Pick one task you usually do manually (researching a topic, drafting a plan, or analyzing data). Use AI as your extension during the research phase, then use the time you save to refine the intent and substance of the final output. Make sure your "hand" is visible in the result.

CHAPTER 2: WHAT AI CAN AND CANNOT DO

In its current form, Generative AI delivers speed and logical output, but it still has clear limitations. It can take on part of the cognitive workload of humans and free up time to think, decide, and create with more intention. Human intelligence, however, is not a single skill. It includes logical reasoning, intuition, critical judgment, emotional understanding, and creativity. To use AI effectively, we need to understand which areas it can support and which still require a human lead, which we will discuss here.

Areas of Human Intelligence AI Can Support

Logical thinking: Generative AI can handle most logical and analytical work. The key is to identify which tasks in your work and life follow clear, logical steps, and then use AI to complete those steps faster and with more consistency. However, AI should be applied by people who understand the fundamentals of the field. Many people focus on AI literacy, but domain knowledge matters just as much.

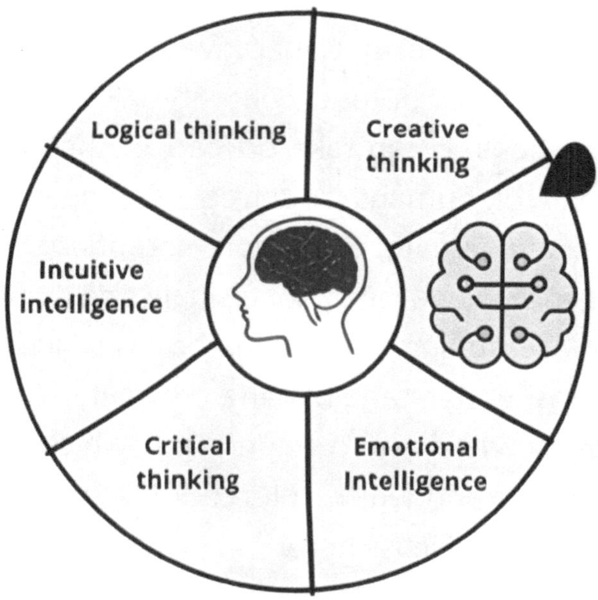

For example, it is unwise to rely on AI for financial decisions without basic knowledge of finance and investing. Without it, you may hand over your critical judgment to whatever the model was trained to predict, instead of making decisions that match your risk tolerance, long-term goals, and lifestyle.

AI literacy without domain literacy is a perfect recipe for blind dependence.

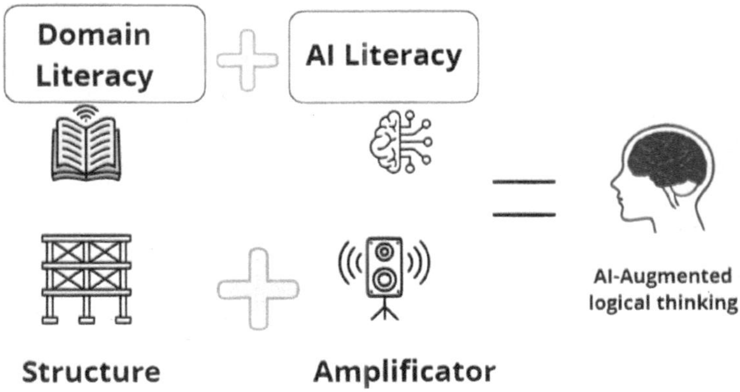

Intuitive intelligence: Intuition is experience-based judgment. It is shaped by patterns we have absorbed over time, often surfacing as a clear sense of direction before we can fully explain the reasoning. Generative AI does not operate through instinct. It produces predictions based on statistical patterns across large datasets. Human intuition remains important when information is incomplete, the context is nuanced, and the decision depends on timing, relationships, or real-world consequences.

Critical thinking: Critical thinking will be our superpower skill in an AI-enabled world. It allows us to evaluate AI outputs rather than accept them at face value. In practical terms, critical thinking is the ability to assess an AI response for context, logic, assumptions, gaps, and potential bias. As Generative AI becomes more fluent and persuasive, we will need stronger discipline to question results before acting on them. One useful way to describe this is a "stop-and-go" mindset: pause to verify the inputs, check the reasoning, and confirm the sources and implications, then proceed only when the response holds up. AI tools can also be used to critique other AI outputs, but the standards for evaluation—what matters, what is acceptable, and what risks are tolerable—must remain human-led.

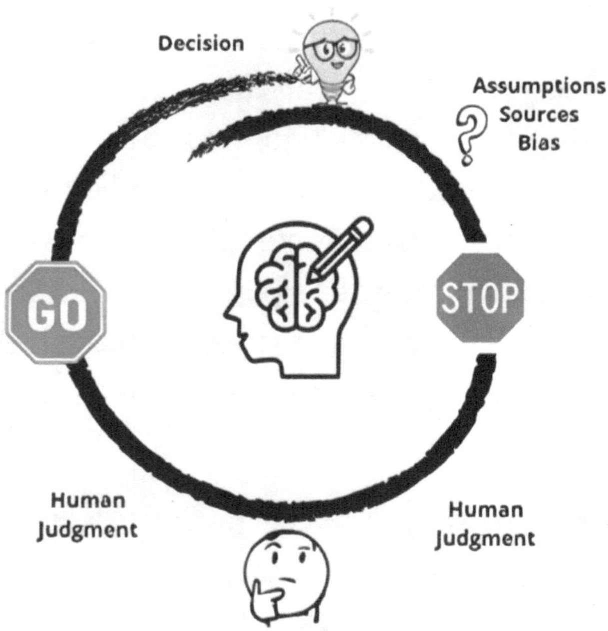

Emotional intelligence: Generative AI can simulate emotional understanding and mimic human emotion, but it does not *feel* emotions or experience the context behind them. It has been trained to identify patterns in language and can respond in ways that sound empathetic, but its responses are based on probabilities learned from data, not lived experience or self-awareness.

Consider a situation where you are frustrated after a dispute with your manager. If you describe the situation to an AI tool, it may suggest responses such as escalating the issue, confronting the person directly, or staying silent. Those options may be reasonable starting points, but they do not address the deeper questions that often determine what a person should do next: Why did this trigger feel so intense? What assumptions am I making? What past experiences are influencing my reaction? What values or needs feel threatened?

In that sense, AI can help you label the emotion, summarize the situation, and generate response options. It cannot, however, replace self-awareness and emotional judgment. You still have to decide what response fits the relationship, the context, and your long-term goals.

Creative thinking: Creativity is often described as an individual skill, but I think of it in two distinct forms. The first form is *combinational creativity*—bringing existing ideas, inputs, and references together to produce a new output. Generative AI can support this type of creativity well. For example, it can help combine themes from multiple documents, generate variations on a message, or propose different ways to structure a piece of content.

The second form is *original creativity*—the ability to introduce something meaningful and new in response to a real need, a specific audience, or a cultural moment. This type of creativity is shaped by lived experience, personal preference, values, and context. AI can imitate styles and produce impressive work that resembles what has come before, but it does not operate from lived context or intention. It generates based on patterns in existing material, while humans create with purpose, awareness of consequences, and an understanding of what is needed now.

A great example of original creativity is the development of fully digital phones—an innovation that did not simply replicate prior designs, but responded to emerging needs and shifting expectations. Similarly, this kind of context-driven creativity produced enduring works such as the *Mona Lisa*, visionary architectural statements such as the Museum of the Future in Dubai, and iconic feats of engineering such as the Eiffel Tower in Paris—each shaped by the constraints, aspirations, and cultural conditions of its time. Some may argue that this form of creativity is merely inspired by different realities; however, while generative AI could produce a *Mona Lisa*–style painting today, it cannot determine what the next *Mona Lisa* of the coming century will be. That answer does not yet exist in the historical record. It will be created by humans who respond to present conditions, anticipate emerging possibilities, and make intentional choices within a particular cultural moment.

In each of the previous areas, we have seen how AI can support human work and add value, but it is clear that human judgment must direct it. Now we need to define its limits. The next section uses the *EVH framework*—Ethics, Values, and Human Judgment—to clarify what must remain human-led.

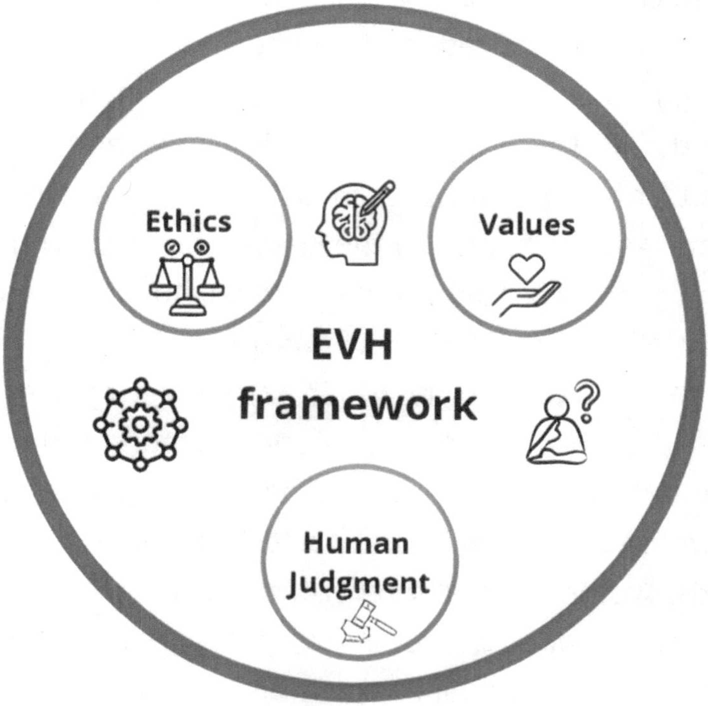
Ethics
Values
EVH
framework
Human
Judgment

What AI Cannot Replace: Ethics, Values, and Human Judgment

When creating any new technology like Generative AI, humans must set standards that determine whether the technology's outputs are appropriate, fair, and responsible. In other words, it is our responsibility to ensure technology serves people, not the other way around. When a tool influences decisions at scale, the non-negotiable requirement is human input—especially in three key areas: ethics, values, and human judgment.

The following questions help evaluate the implications of a new technology on our human values:

1. Are universal human values kept at the center of the technology's development?

2. Are we proactively assessing the impact of this technology on our ethical system?

3. Are we maintaining human control over judgment and decision-making?

If the answer to all three questions is yes, it signals a responsible intent to develop a system that can be used for good to create significant value for both current and future generations.

Can an AI system alter these three elements? Some AI engineers might say yes, but this is the boundary every person using Generative AI should set to adopt the technology responsibly, thus protecting people from misuse and manipulation and reinforcing fairness and equity.

When I say AI cannot replace our system of values and ethics, I mean that humans must remain the ones who decide what is acceptable, what is not, and what ethical use looks like in practice.

For example, if a school permits tools like ChatGPT to improve learning, the goal should be to integrate them into the classroom's ethical standards and guidelines. That approach promotes responsible use rather than banning students from a powerful learning tool.

In short, Generative AI can support execution, but it cannot set its own values, apply its own ethics, or make its own judgments. Those responsibilities require human leadership. The EVH framework is the practical checkpoint that keeps AI useful, accountable, and aligned with the people it affects.

Authentic Communication: Keeping the Human in the Message

The final thing AI cannot do is to become the authentic voice of humanity. As Generative AI becomes part of everyday writing, the risk is not that messages will become "bad." The risk is that they will become **generic**—polished on the surface, unclear in intent, and disconnected from the person responsible for them. AI can mimic various familiar voices, but it won't create authentic voices AI can draft and refine quickly, but it cannot fully understand context, relationships, power dynamics, or consequences. When people send AI-written messages without review, they may communicate something they do not actually mean, create confusion, or damage trust.

A common problem shows up in routine workplace communication. Imagine a manager asks an AI tool to write a response to an employee who missed a deadline. The AI produces a firm, professional email that sounds reasonable. But the manager fails to adjust for the real context: the employee has been covering extra shifts, the deadline was changed late, and the employee is already anxious about job security. A message that reads "clear and direct" on the screen can land as punitive, dismissive, or unfair in real life. The tool did its job—generate text. The human skipped theirs—apply judgment.

This is the practical standard for authentic communication: use AI as an assistant, not as your speaker. Let it help you organize your thoughts, tighten language, and reduce time spent drafting. Then take control of the final message. Ask: Does this reflect my intent? Does it fit the relationship and the situation? Would I stand behind it if it were read aloud in a meeting? That final human review is where responsibility lives. It is also how you ensure technology serves people, not the other way around.

Each of us has been growing up, living, and working in different sets of environments, cultures, and countries and all of this would have shaped our deep identity, and that's how our thoughts are shaped uniquely based on our life experiences. If AI decides to step into this space, this will harm our own identity. That's why I believe that each of us should decide intentionally to not give away our inner voice to AI.

Summary

The Core Shift

- **The old way**: Treating AI output as a substitute for domain knowledge and critical review.

- **The new way**: Using AI for speed and structure while keeping human leadership in judgment, context, and accountability.

Key Takeaways

- **Domain literacy remains essential**: AI can accelerate work, but it cannot replace foundational knowledge in the field where it is applied.

- **The "stop-and-go" mindset protects decision quality**: Pause to test assumptions, verify inputs, and evaluate implications before acting on AI output.

- **EVH is the boundary**: Ethics, values, and human judgment must remain human-led to ensure technology serves people, not the other way around.

- **Authentic communication is a responsibility**: AI can draft, but humans must ensure intent, tone, and consequences are appropriate to the situation.

The Reader's Audit

Answer each question below:

- **Delegation map**: In my current work or studies, what tasks are repeatable and logic-based (research summaries, comparisons, calculations, first drafts) that AI can support? What tasks require human leadership (judgment, context, priorities, and accountability)?
- **Domain check**: Where am I using AI in a field where my foundational knowledge is weak? What is one core concept I need to understand so I can evaluate AI outputs responsibly?
- **Stop-and-go check**: The last time I used AI, what assumptions did I accept without testing? What would I verify next time (inputs, sources, constraints, risks, stakeholders)?

- **EVH checkpoint**: For one real use case in my life (school, work, or personal), can I answer "yes" to all three EVH questions?
    - Values centered
    - Ethical impact assessed
    - Human judgment remains in control
- **Communication check**: Where do I send messages that carry real consequences (emails, texts, performance feedback, academic submissions)? What will my review standard be before I send or submit AI-assisted writing?

The Next Step

Tomorrow morning, choose one task you normally do manually (researching a topic, drafting a plan, or analyzing data).

- **Use AI for the first pass** (summarize, compare options, outline, calculate, or generate a draft).

- **Apply "stop-and-go" before you act**: verify key assumptions, confirm constraints, and test the output against your context.
- **Run EVH**: check values, ethics, and whether you remain accountable for the decision.
- **Finalize the human layer**: revise the output so the intent, tone, and decision are clearly yours.

CHAPTER 3:
KNOWLEDGE, EXPERTISE, AND INTELLIGENCE

For much of modern history, social status and professional advantage have been tied to knowledge. Now, we are entering the AI era, where access to knowledge and expertise is being democratized, leaving smaller gaps across the workforce.

For decades, we have confused "expertise" with "memory." We rewarded the person who could recall the most information. But AI is turning knowledge into a utility, like electricity. When I plug in electricity, the bulb lights up the street. When I plug in AI, expertise becomes available on demand. We do not store electricity inside the bulb, and in the same way, we will not need to "store" expertise in our heads to prove our value. Today, being an expert is less about what you remember and more about what you can architect using the tools at your fingertips.

Some professionals have built twenty years of expertise in a specific field, and now, Generative AI can sometimes produce faster and more accurate outputs in routine cases than they can. After social media emerged in the early 2000s, personal brand often became more visible than knowledge itself. In the AI era, I believe authenticity will matter more. If anyone can talk about anything, then who “owns” knowledge? Is it ChatGPT and its massive datasets and neural networks? In the past, deeply specialized knowledge was often limited to a small group of experts—PhD holders who built theories, and academics at top universities. In the future, expertise will often be materialized through AI systems that can deliver high-quality answers at scale. This means we must reinvent how society recognizes and rewards knowledge and expertise.

In that context, imperfection becomes a proof of humanity. Human experts will still matter because we can be wrong, be vulnerable, and have “skin in the game” in a way an AI system cannot. The expert’s value will increasingly be demonstrated through judgment, accountability, and decision quality—not only through technical output.

A clear example is enterprise technology. Consider a professional who built expertise configuring enterprise resource planning (ERP) systems, such as SAP. In the past, they had to know the platform's rules, settings, and process logic in detail. In the AI era, Generative AI can assist with parts of that work—drafting configuration options, generating documentation, producing test scripts, and accelerating troubleshooting. As a result, the expert's role shifts upward. The expert becomes responsible for defining requirements, validating outputs in the real environment, identifying failure modes, protecting controls and compliance, and translating technical choices into business impact for stakeholders. In other words, AI can support configuration, but the expert must lead judgment, governance, and accountability.

In the future, we will not need to know everything. We will need to use what we know better than others. For example, if I have an autonomous AI agent that can run financial analysis on every company in the world, I can be viewed as a top expert if I can turn that analysis into better investment decisions than others. The value comes from how I use, train, and improve the agent's outputs and how I apply judgment based on risk tolerance, goals, and real constraints. I do not need a PhD in finance to create value in this environment. The new credential is the ability to synthesize, test, and apply AI-assisted insights responsibly—what I call a “Master of Experimental Synthesis.”

At the same time, we still need ways to recognize experts through verifiable credentials and demonstrated competence. If an expert can show how they used AI and human judgment to make a decision—what assumptions they tested, what risks they considered, and what tradeoffs they accepted—then that person should receive credit for the AI-augmented expertise and remain accountable for the outcome.

I also believe this shift will accelerate the rise of the *polymath*—a person with extensive knowledge and expertise across multiple, often diverse fields, with meaningful contributions in more than one domain. Polymath comes from Greek words for "much learning," not just mathematics. Famous examples from the past include Leonardo da Vinci, who excelled as an artist, engineer, and inventor. In the future, a professional who can use AI agents in both finance and art could create a niche field that helps middle-income investors identify promising art investments, potentially expanding access to art-market decisions that have traditionally been limited to wealthy collectors and insiders.

As we define new values for expertise and knowledge in the era of AI, we also need new reference points for intelligence. In the past, intelligence was often defined primarily through IQ (intelligence quotient), which measures abilities such as reasoning, analytical thinking, working memory, processing speed, and verbal and mathematical skills. IQ became a fundamental way to benchmark intelligence and was often correlated with academic performance. However, AI has intelligence similar to what is measured on an IQ test. For that reason, we need to evolve our assessments of intelligence to recognize other forms of intelligence, such as emotional intelligence, creative intelligence, cultural intelligence, social intelligence, moral intelligence, intuitive intelligence, spiritual intelligence, practical intelligence, and more.

Most of these forms of intelligence are developed through life experience, and many are not fully recognized in educational systems today. As knowledge is democratized, these non-IQ intelligences will increasingly become part of how we assess capability in both the workforce and education. Historically, these intelligences were often treated separately across disciplines—for instance, creative intelligence was mostly associated with the arts. However, many jobs of the future will require advanced creativity to stay ahead of machine automation.

As machines automate repetitive work, intelligence will be measured more by our ability to elevate the work—adding interpretation, meaning, and direction. As AI capabilities expand, human value concentrates in what must remain human-led: context, judgment, and responsibility for decisions.

AI will be able to communicate artificial emotions, build artificial empathy, and navigate cultural nuances artificially, but it will not articulate the genuine connection across all with our own human intelligence.

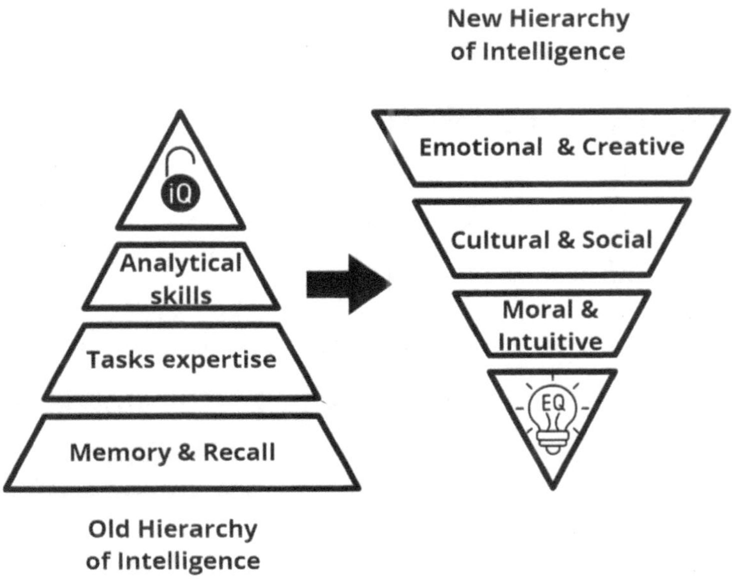

Summary

The Core Shift

- **The old way**: Treating expertise as memory and rewarding recall.
- **The new way**: Treating knowledge as a utility and defining expertise by how well you synthesize information, apply judgment, and create value with AI support ("Master of Experimental Synthesis").

Key Takeaways

- **Knowledge as utility**: As access to information expands, expertise is demonstrated through application—how you use tools, not how much you can memorize.
- **Modern expertise**: The expert's role shifts from producing answers to orchestrating tools, validating outputs, identifying risks, and translating insights into decisions.
- **The new polymath**: AI lowers barriers between disciplines, creating opportunities for professionals who can bridge fields (for example, finance and art) to build new niches.
- **Beyond IQ**: As AI matches analytical capabilities, differentiators include practical, cultural, moral, and intuitive intelligence—especially when decisions require context and accountability.
- **Skin in the game**: Human value remains in responsibility for outcomes—owning trade-offs, learning from error, and standing behind decisions informed by AI.

The Reader's Audit

If you no longer need to "know everything," what niche could you develop by combining AI with one domain you know well and one area you are interested in?

In your work or studies, where does AI's analysis end and your judgment begin? Identify one decision you will keep explicitly human-led.

The Next Step

This week: Choose one AI-generated output you would normally accept at face value (a summary, recommendation, ranking, or plan). Improve it using your domain knowledge and practical context. Document what you changed and why, and then make a decision you can defend.

CHAPTER 4:
THE PURPOSE CRISIS

We are now entering what I call the "Purpose Crisis." For a century, many of us defined ourselves by our productivity. However, in the AI future, we will move from a utility-based purpose ("I am what I do") to an essence-based purpose ("I am why I do it") as Generative AI takes on more of our tasks and helps us reclaim our most finite resource: time.

When this happens, many people may experience a loss of purpose and may need to reassess what gives their work and life meaning. Purpose is one of the most difficult questions people wrestle with across a lifetime, and this shift caused by Generative AI forces it into the foreground for anyone whose identity has been anchored mainly in what they produce. The goal is not to abandon your skills, but to expand your identity beyond them and clarify what you want to contribute in an AI-enabled world.

Time as the Real Disruption

When we gain time, we gain choices. The AI era will push many people to ask a new set of questions: What do I want my work to mean? What do I want my life to be about when routine output is no longer the main measure of value? What do I want to build, learn, or give that cannot be automated?

This is why the purpose crisis is not only about jobs. It is about identity. If your sense of worth has been anchored mostly in productivity, then reclaimed time can feel liberating, but it can also feel disorienting. Time creates space. Space forces priorities.

The Life Energy Equation

To assist you with this shift, I have developed a personal model that has guided my understanding of purpose and happiness. I call it the **Life Energy Equation**:

Life Energy = (how much you love your passions) × (your passions) × (time spent nurturing passions and interests - time spent away from them)

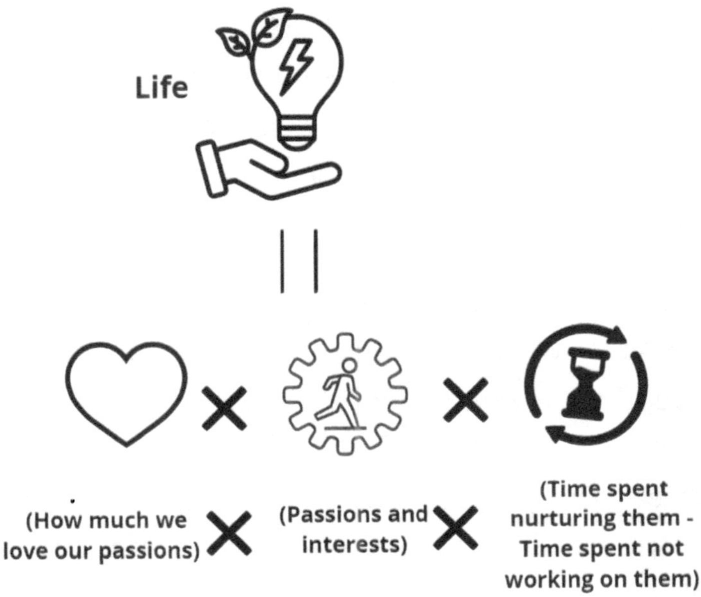

Let me explain my theory. Life is like a battery—a finite supply of energy—and many people believe that energy is given by God to fulfill a mission. When you spend time on what you love, you are charging. When you invest most of your energy in work that can be automated, you are leaking. Over time, our energy is maximized when we consistently nurture our passions and interests. That is often where a deeper sense of fulfillment comes from, because it aligns our daily effort with our larger purpose.

On the other hand, the more time we spend away from our passions and interests, the lower our energy tends to become. This misalignment can contribute to feeling stuck, burned out, or disconnected from purpose. In some cases, it may help explain why people experience prolonged discouragement or low mood when their work and their values stay misaligned for too long.

This is where Generative AI changes the equation in a positive way. AI can give time back by reducing repetitive, low-value work. That reclaimed time creates space for self-reflection—space to better understand yourself and clarify what you want to contribute. For some people, that may lead to community involvement. For others, it may lead to retraining, career reconversion, or a completely new chapter of life.

If we refocus our time on missions that complement our work, we can find a "sweet spot." In that space, AI supports what we do not value as much—routine, repetitive tasks—while we keep our focus on human-centered work that machines cannot substitute: judgment, creativity, relationship, service, and meaning.

Many people benefit from guidance during this kind of transition. I have benefited from coaching, and I also coach others in their growth journey. At its best, coaching is practical: it helps people clarify priorities, test assumptions, and make decisions they can follow through on. As AI takes over parts of routine work, that kind of guidance can help people use reclaimed time with more intention and purpose.

Summary

The Core Shift

- **The old way**: Defining our identity through "utility-based purpose," the belief that, "I am what I do," and that my value is tied to my technical productivity.
- **The new way**: Transitioning to an "essence-based purpose" and realizing that, "I am why I do it"—using AI to reclaim our time and refocus on our soul's mission.

Key Takeaways

- **Reclaiming time**: AI is the mechanism that allows us to recover our most precious resource: time. It is the invitation to move from handcrafted survival to a life of intentional self-reflection.

- **The Life Energy Equation**: Life is a battery. When we nurture our passions, we are charging. When we spend our energy on tasks that can be automated, we are leaking. Happiness is found in the "sweet spot" where AI handles the leakage and we handle the charge.

- **Human questions**: In a world overflowing with artificial answers, the true genius lies in the art of the question. Coaching becomes the essential growth industry because it supports the human "hand" in finding its direction.

- **The mission**: As machines take over parts of our work, we must take over new parts of our passions. We must use the time AI gives back to discover what we can uniquely give to humanity.

The Reader's Audit

Using the Life Energy Equation, identify one area of your work that is currently a "leak" in your battery. How can you use an AI extension to stop that leak?

If you reclaim your time through AI, what is the human-centric mission or passion you want to charge? What does your "essence" want to create?

The Next Step

This weekend: Stop being the "provider of answers" for a moment. Practice the art of the question with yourself or a friend. Reflect on this: "What part of my work brings me energy, and what part is just a habit of utility?" Use your pen to capture the mission that makes you feel most alive.

CHAPTER 5: SURVIVAL SKILLS GUIDE

Generative AI is bringing a new set of skills into every job, and two skill areas will matter across nearly every discipline: *AI engineering* (building and maintaining AI solutions) and *AI literacy* (using AI tools effectively). In the short term, AI literacy is the most urgent baseline for most people. *AI literacy* means knowing how to use Generative AI in a smart, safe, and practical way. It includes prompt engineering (writing clear instructions), using AI agents to support business processes, and applying responsible-use principles.

Over the next five to ten years, AI literacy will become a form of digital literacy. Just as it is now intuitive to use a phone and Google to find a library or a shopping mall, it will become intuitive to use AI tools and agents to solve everyday work and learning tasks. This is why we are moving from the Age of Keyboards to the Age of Intent—where results depend less on typing and more on clearly expressing what you want, why you want it, and what constraints matter.

This shift also lowers the barrier to entry. To use AI daily, many people only need a mobile device, an AI application, a microphone, and an internet connection. In many cases, they can speak as they work. Compared to earlier digital literacy—where a laptop and basic IT skills were often required—AI can be easier to access and faster to apply. However, that accessibility creates a parallel risk: the gap will widen between those who learn to use AI effectively and those who do not. The risk is multigenerational, and it is practical, not theoretical. If AI literacy becomes a baseline for opportunity, then expanding access becomes a shared responsibility so people are not excluded from education, work, or advancement.

Why Technical Skills Are Shifting

Every technical skill, including coding, has an expiry date and will face increasing pressure to evolve. AI is already automating many tasks that were traditionally performed at the junior level of jobs. Whether it is writing an email, designing a presentation, or drafting strategies and plans, AI systems can often do more with fewer human hours. This is transforming the skillsets required to perform many roles.

Even if we do not change our jobs, our jobs will change around us. Human skills will be reprioritized. Skills such as critical thinking, communication, leadership, and conflict management will become more central because they sit closer to judgment, relationships, and decision-making. Machines will automate tasks, and humans will increasingly focus on the human side of the work.

The 4-Quadrant Skills Filter

The skills that are most likely to be replaced, augmented, or left largely untouched in the AI era, can be organized into four categories:

1) Technical skills with low complexity

These tasks have a higher risk of being replaced or heavily automated. For example, creating a basic website can now be done with a prompt.

2) Technical skills with high complexity

These tasks will be augmented by AI, but they are less likely to be fully replaced because they involve deeper architecture, integration, and problem-solving. For example, building a complex website still requires advanced technical judgment, design choices, and experienced oversight—often supported by coding agents.

3) Human skills with low-stakes interaction

These tasks have a higher risk of being replaced or automated because the interaction is routine and the consequences are limited. For example, a call center role that follows a simple troubleshooting script may be handled smoothly by an AI-based digital assistant.

4) Human skills with high-stakes interaction

These skills are more resistant to automation because they involve real consequences, trust, negotiation, and accountability. A strong example is contract negotiation, where context, risk, and human judgment matter.

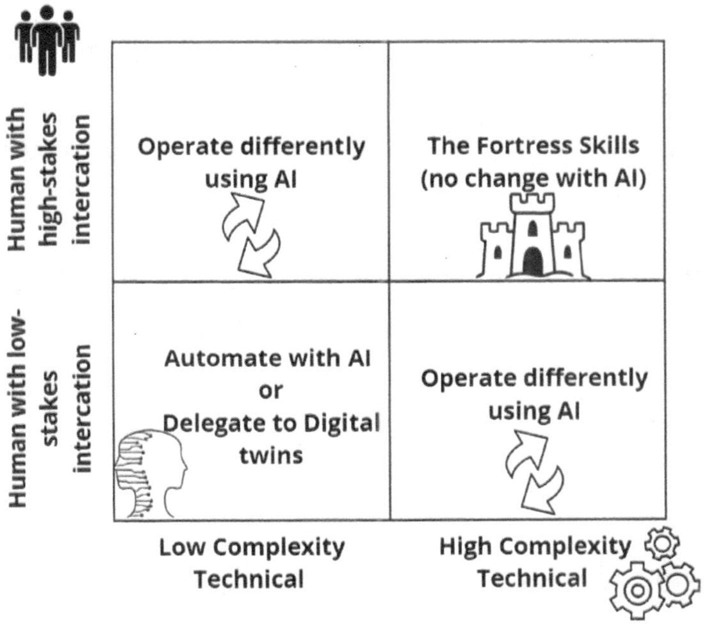

The Ultimate Muscle: Adaptability

AI can also be trained on procedures and processes related to specific skills, and it will become more accurate over time. In some areas, it may outperform humans on speed and consistency. This makes adaptability essential. We will need the discipline to continue learning throughout long careers.

For that reason, transferable skills become a form of career stability. Skills such as communication, problem solving, leadership, and decision-making can be applied across technical and non-technical roles. They also provide flexibility to shift roles within an organization—or to move toward entrepreneurship by thinking about career growth through skills, not titles.

In an era of rapid change, the ultimate skill is the ability to unlearn, relearn, and learn new behaviors and competencies. Agility becomes a practical form of job security.

A Personal Reality Check

When I entered the workforce more than ten years ago, I could not imagine that much of the work I did as a junior consultant would be automated today—analyzing documents, collecting information, doing research, and analyzing data. If earlier shifts took years to become automated, upcoming shifts may happen in months. The future belongs to those who invest in learning early and maintain a growth mindset that keeps them relevant as roles evolve.

Summary

The Core Shift

- **The old way**: Operating in the "Age of Keyboards," where value was tied to technical IT literacy, manual research, and the mastery of junior-level tasks like data collection.
- **The new way**: Entering the "Age of Intent," where value is tied to AI literacy and the ability to orchestrate agents through voice and instruction to solve complex problems.

Key Takeaways

- **AI literacy as a human right**: Just as Google became our intuitive search tool, AI agents will become our intuitive life partners. We have a collective responsibility to ensure no generation is left behind in this transition.
- **The expiry date of technical skills**: Even if we don't change our jobs, our jobs will change around us.

AI is already automating the "junior" layers of work such as writing, drafting, and researching, forcing us to move higher up the value chain.

- **The 4-Quadrant Filter**: Not all skills are created equal. High-stakes human interactions and complex technical architecture remain our safe harbor, while low-complexity and low-stakes tasks are rapidly being automated.
- **The ultimate muscle**: In an era of rapid shifts, the most important competency is the ability to unlearn, relearn, and learn. Agility is the only true job security.

The Reader's Audit

Reflecting on your daily routine: Which of your tasks fall into the "Low-Complexity Technical" or "Low-Stakes Human" categories? These are your "vulnerability zones"—how will you start delegating them to AI today?

What is one "High-Stakes Human Interaction" (like a difficult negotiation or a leadership challenge) that you have been avoiding? How can you use the time saved by AI to master this "immune" skill?

The Next Step

This Week: Move from a "Keyboard" mindset to an "Intent" mindset. Use the microphone on your AI application to "speak your intent" for a complex task. Don't just ask for a result; describe the context, rationale, and desired soul of the output. Observe how your "hand" shapes the machine's "extension."

CHAPTER 6:
JOBS OF THE FUTURE

Many of the jobs of the future do not even exist today, and in five to ten years, many of today's jobs will be replaced by machines. These are most likely roles where limited human interaction is needed, or where human intervention can be replaced by training an AI system to perform the work with speed and consistency.

A simple example is basic translation. When AI models are trained and incorporated responsibly—with transparency and bias controls—many routine translation tasks can be completed faster and, in many cases, more consistently by AI than by humans. By contrast, jobs that rely heavily on human skills, such as sales, will be less impacted by AI replacement and more likely augmented. These roles are not easily replaced because the work depends on trust, judgment, and human decision-making in real time.

If you are already in the workforce, I recommend assessing how your job will evolve in the AI era using the BCP (Break Down, Classify, Plan) framework below.

The BCP Framework (Break Down, Classify, Plan)

Step 1: Break Down

Break down the skills and tasks required in your job.

Step 2: Classify

Classify the tasks you do into four categories, and then rank them by how likely they are to be automated or reshaped by AI:

A. Routine and replaceable by AI

B. Routine but complex (requires deeper cognitive, contextual, or creative work)

C. Human skills with low complexity in execution

D. Human skills with high complexity in execution ("fortress skills")

Step 3: Plan

Plan your upskilling journey now. Do not wait.

To show how this works in practice, I will use my current job as a Customer Success Manager and walk through a full analysis in the section below.

BCP framework

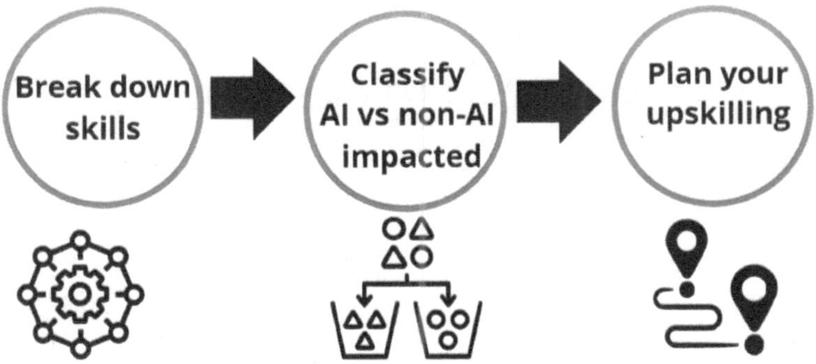

Example: Applying BCP to a Customer Success Manager Role

Step 1: Break Down Skills and Tasks

I analyzed a typical week of work to identify the tasks and skills that are critical to performing my job.

The list below reflects the areas that most consistently drive outcomes:

- Building presentations
- Sending customer communications
- Training customers
- Gathering customer health data
- Interpreting client data within a specific context
- Coordinating subject matter experts
- Storytelling
- Prioritizing clients based on defined criteria
- Strategic thinking
- Conducting basic advisory meetings
- Conducting strategic alignment meetings
- Negotiation
- Crisis management
- Leadership

Step 2: Classify Skills into Categories

A. Routine tasks that can be automated by AI

Sending communications, building presentations, organizing meetings, prioritizing clients based on defined criteria, and gathering customer data.

B. Routine tasks that require complex cognitive, contextual, or creative work

Analyzing health data, coordinating experts on routine topics, and interpreting client data within a specific context.

C. Human skills with low complexity in execution

Training customers and conducting basic advisory meetings.

D. Human skills with high complexity in execution ("fortress skills")

Conducting strategic alignment meetings, storytelling, negotiation, strategic thinking, crisis management, and leadership.

Step 3: Planning My Upskilling Journey

My plan follows directly from the four categories:

For A (routine and automated): I will strengthen my AI literacy so I can use AI tools and agents to automate routine execution reliably.

For B (routine and complex): I will strengthen my ability to audit AI outputs using critical thinking, context checks, and validation against real customer constraints.

For C (human and low complexity): I will use AI to improve preparation and follow-up so I can deliver these interactions more efficiently while keeping the engagement human-led.

For D (fortress skills): I will build a deliberate plan to strengthen these skills because they are transferable across roles and remain valuable in any market. For example, to improve at negotiation, I need to practice empathy and active listening.

The key thing to understand as we move toward the jobs of the future is that no company can map your role fast enough to keep pace with change. Do not wait for a new job description. Use the BCP framework to write your own.

When you audit your tasks today, you move from being a passenger in the AI revolution to being the pilot.

Summary

The Core Shift

- **The old way:** Waiting for a company to redefine your role, while relying on traditional expertise and "knowledge retention" as your primary form of job security.
- **The new way:** Becoming the pilot of your own career by auditing your tasks and using the BCP framework to stay ahead of automation.

Key Takeaways

- **The BCP Framework:** Break down your tasks, classify them by complexity and human judgment, and plan your upskilling before change is forced on you.

- **The fortress skills:** High-stakes human skills—storytelling, negotiation, crisis management, and leadership—are the most resistant to automation because they require judgment, trust, and accountability.
- **The automation ladder:** Routine junior-level tasks are often the first to be delegated. If you are not learning how to automate or redesign them, you are falling behind.
- **Ownership over occupation:** Your job will change even if your title does not. Professionals who use AI as an efficiency tool can redirect time toward higher-impact human work.

The Reader's Audit

1. List your top ten to fifteen tasks. Which ones are Category A (routine and replaceable)? Which are Category D (fortress skills)?
2. What is one fortress skill you need to strengthen this quarter, and what is one routine task you will automate or redesign to make time for it?

The Next Step

This week, run a BCP audit on your role. Don't wait for permission. Automate or redesign one Category A task, and then invest the time you gain into one Category D skill through deliberate practice, feedback, or training. Use the "Hand" of your intent to review it, and notice how much "Time and Power" you get back for your "Fortress Skills."

CHAPTER 7: EMPOWERED LEADERSHIP: BUILDING AGILE HUMAN–AI TEAMS

AI is redefining not only skills and jobs, but also organizational structure and leadership by democratizing access to information inside the organization. As information becomes easier to access and share, vertical specialization will no longer be a reliable indicator of leadership success. While many of us have seen "irreplaceable" technical gurus rise through the ranks because they retained critical knowledge and mastered expertise that few others could reach, in the job market of the future, that gatekeeping advantage will matter less. Organizations will move from highly hierarchical models to flatter ones, and from information retention to orchestration—allowing expertise to flow more freely, both horizontally across teams and vertically through internal mobility.

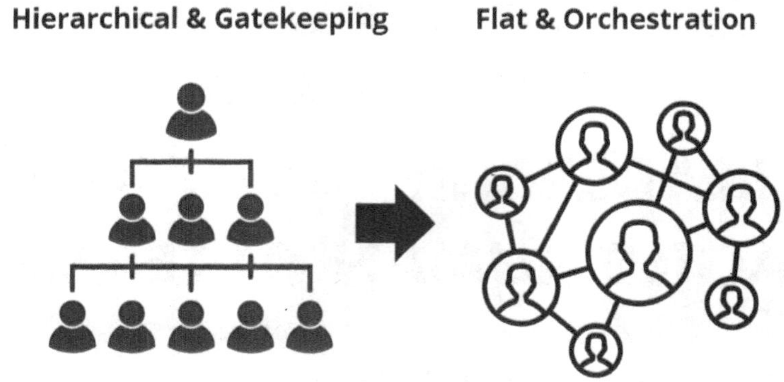

Teams of the Future

Teams of the future will be composed of humans and AI agents. Middle management will become more critical, not less, as managers take on responsibility for responsible AI use, trust, and accountability within teams. At the same time, middle management will need less administrative complexity and more leadership clarity. Managers of the future will operate more like coaches and change leaders. Their effectiveness will be less tied to the technical track they come from and more tied to their ability to federate teams of humans and AI agents around shared priorities and standards.

This shift will not be limited to managers. Executives will increasingly focus on direction, governance, and trade-offs—setting the questions, constraints, and decision standards that shape how AI is used at scale. Technical leaders will move from being the primary source of answers to being architects of systems—designing workflows, validating outputs, and ensuring reliability and security. Frontline supervisors will focus on adoption and execution—helping people integrate AI into daily work while maintaining quality and accountability. Project leaders and team leads will act as orchestrators—coordinating human and AI contributions, assigning work to the right "doer," and keeping outcomes aligned with business goals. HR leaders will play a larger role in capability building—defining AI literacy expectations, supporting upskilling, and reinforcing ethical norms. Across functions, the common thread is the same: as AI increases speed and access to information, leadership becomes less about controlling knowledge and more about building alignment, judgment, and trust.

The AI agent will become like a team member, creating a shared space for collaboration between people and tools. An AI agent only creates value when it is part of the everyday workflow, not an occasional tool someone tries once in a while. The team has to train it, test it, and improve it over time so it reflects real processes and real constraints. That requires information sharing—people have to document what works, flag what fails, and surface risks early. This is where management matters: leaders must create an environment where teams can challenge AI outputs, report mistakes, and adjust processes without blame. In this context, AI will provide the breadth of information, but the human will provide the depth of accountability. The most performing teams of the future will be creating their own AI agents that can automate the specifics of their own jobs and markets. To reach that high level of AI fluency within the team, incentives need to be created in advance to encourage the use of Generative AI, information sharing, and to reward the innovators of the team.

AI also cannot be held responsible for building an agile culture that can succeed in the AI era. It will not maintain trust, resolve conflict, or carry accountability when decisions accelerate. AI can generate options and scenarios, but it cannot replace judgment. It can propose a vision, but it cannot do the human work required to align people behind it.

Leadership, then, becomes less about controlling information and more about framing the right questions, setting clear governance, and empowering teams to think critically. Authority will not be tied as closely to knowledge asymmetry. It will be tied to empowerment—creating systems where employees can challenge assumptions, test scenarios, and propose alternatives. If teams are not empowered to challenge leadership decisions, they will not be equipped to challenge AI outputs either.

Imagine a decision made by a leader based on AI recommendations, but the decision lacks a perspective that is critical to business success. The team must be empowered to surface what was missed and to propose better options, including using AI to test alternatives. That is the kind of operating system that helps reduce misuse—by leaders and employees alike—and creates a culture of accountability and critical thinking. Leaders should define governance around AI use while encouraging collaboration, transparency, and disciplined evaluation.

Future leadership will be all about humility, adaptability, judgment, and authenticity. In an era where access to technical skills, information, and expertise will be democratized. AI may create the most effective and efficient strategic plans, even without the help of big consultancies. However, AI will not build an agile culture that can succeed in the AI era. AI will not maintain trust when directions to every problem accelerates decisions, and it won't replace judgment neither. AI can create a vision, but it cannot inspire. Leaders will be the ones who know how to frame the questions that AI can answer. Authority won't be linked necessarily to knowledge asymmetry. Leadership will all be about empowerment, rather than a one sided and one directional conventional system that rewards one way of thinking.

Agile Organizations

Organizations of the future will look more like networks than vertical hierarchies. If Generative AI reduces information asymmetry, then some middle-management work that existed primarily to cascade information and translate strategy may be needed less in its current form. In parallel, organizations may build centralized agents—such as a "People GPT"—to synchronize access to internal knowledge. In that environment, the critical lever becomes participation: leaders must create incentives for people to feed the system with timely updates so the AI remains current and useful.

Imagine a team of ten employees with access to multiple agents. The manager's role shifts away from information cascade and coordination meetings and toward accountability, belonging, conflict management, and skill development. Authority increasingly comes from leading by example and managing the human dynamics of work. In practice, the manager often decides whether a task or project should be done, while the team and AI determine how to execute it.

As we depend less on one another for information, we risk weakening the social tissue that holds a company together. The manager of the future becomes the architect of belonging, ensuring that in a world of high-tech efficiency, teams do not lose the collaborative spirit and human connection that make organizations resilient.

Summary

The Core Shift

- **The old way:** Leadership based on knowledge asymmetry and vertical gatekeeping, where power came from retaining information and moving it slowly down a hierarchy.

- **The new way:** Leadership based on orchestration and empowerment, where information is democratized and the leader's value is found in building trust, accountability, and belonging.

Key Takeaways

- **From gatekeeper to orchestrator:** As AI reduces information gaps, the "technical guru" model weakens. The modern leader federates teams of humans and AI agents, clarifies priorities, and builds accountability while teams and tools improve execution.

- **The "People GPT" network:** Organizations will shift from vertical towers to horizontal networks. Centralized agents can synchronize knowledge, reducing the need for traditional information cascade.
- **The architect of belonging:** As teams rely less on one another for data, organizations risk losing social cohesion. Leaders must ensure efficiency does not erode collaboration and trust.
- **Authority through vulnerability:** In an era of highly polished AI outputs, a leader's authority comes from humility, judgment, and the ability to be challenged. If a team cannot challenge its leader, it cannot effectively challenge AI results.

The Reader's Audit

- **Gatekeeper vs. orchestrator:** In your leadership style (or your manager's), what percentage of influence is derived from information retention? What would it look like to shift that toward information sharing this week?

- **The social tissue:** If your team can get answers from AI without speaking to colleagues, what specific human ritual will you introduce this week to strengthen belonging?

The Next Step

This week: Run a "scenario challenge." Take an AI-generated strategic plan and present it to your team. Explicitly invite them to identify what the plan misses from a human, operational, or stakeholder perspective. Reward the most useful critique and the best alternative. This is how you build an agile culture of trust and accountability.

CHAPTER 8: EDUCATION IN THE FUTURE

Education is the cornerstone of any upskilling journey. For centuries, institutions have brought people together to learn, test ideas, and build what advances society. That role will not disappear in an AI-first world. What will change is how education prepares students for a labor market where many entry-level tasks are being automated and where AI literacy is becoming a baseline expectation.

AI will not replace education's fundamental role in building judgment, instilling values, and preparing the next generation to contribute responsibly. However, AI is already reshaping early-career work. In several sectors, particularly technology and consulting, organizations are using AI tools and agents to automate tasks that used to be assigned to interns and junior staff—research, drafting, data collection, and basic analysis. As that trend grows, the expectation for new graduates will shift: students will need to know how to use Generative AI across common business and learning workflows, while still demonstrating real understanding of the discipline.

Some jobs of the future do not exist today, and many jobs of today will be fully or partially automated. Universities, therefore, must adapt. One practical way to view this shift is through four angles:

1. Teaching with AI
2. Learning with AI
3. Assessing with AI
4. Employability with AI

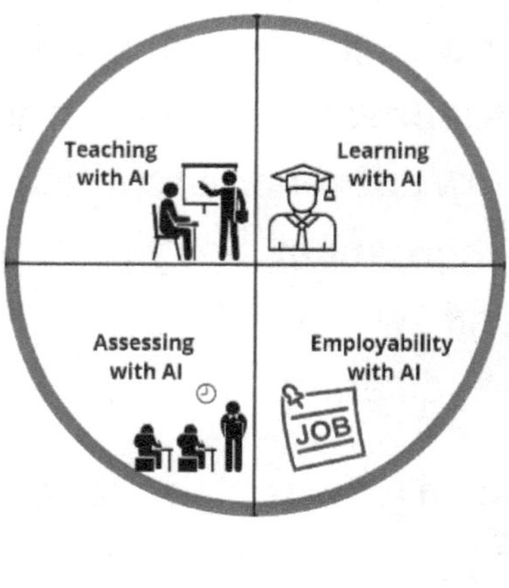

1. Teaching with AI

Professors are a critical pillar of the AI transformation of schools and institutions. They are role models, and their choices influence student norms.

If faculty treat AI as a credible tool—used responsibly and transparently—students are more likely to adopt it with discipline. At the same time, it is understandable that some educators may resist change if they perceive AI as a threat to their expertise or their role.

AI tools can already solve complex math, physics, and multidisciplinary problems quickly. That does not eliminate the need for teachers; it changes what teaching must emphasize. Professors still anchor foundational concepts, but they increasingly need to complement instruction with experiments, projects, and real-world applications that include Generative AI tools.

In that sense, the professor of the AI future moves away from the "sage on the stage" model and toward a "coach who guides the tribe"—a curator of intellect and a lab manager for human–machine collaboration. The classroom becomes less about delivering answers and more about building the habits that produce good judgment: asking better questions, testing assumptions, documenting reasoning, and improving outputs through iteration.

A useful analogy is the evolution of driving. Early car owners needed to be amateur mechanics just to operate the machine. Over time, cars became more automated, and what people needed most was not mechanical skill, but competence in navigation, safety, and responsible decision-making. Education faces a similar shift. As AI makes "calculation" easier, teaching must focus more on "navigation"—how to use AI responsibly, how to recognize bias and limitations, and how to coordinate work across humans and AI tools in complex situations.

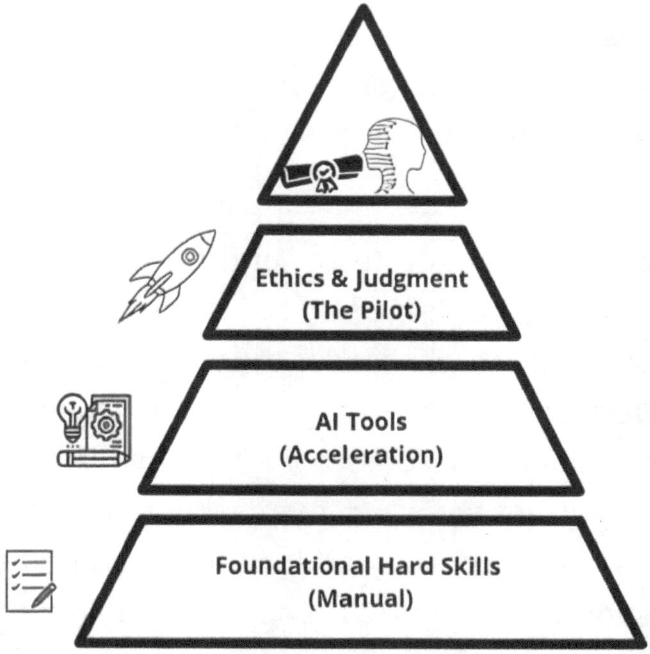

"AI Literacy Stack" for students

2. Learning with AI

Students must learn what to delegate to AI and what not to delegate. The boundary depends on the purpose of the course and the skills being assessed.

Course objective: If the goal is to solve a complex math problem, students should still build the thinking process: how they structure the problem, how they select methods, and how they validate results. AI can accelerate parts of the work—for example, solving sub-problems or checking calculations—but the student should document how AI was used and how the output was evaluated.

Skills assessed: For soft skills such as communication, AI can serve as a tutor and practice partner. A student preparing a presentation can draft a first version, then role-play with AI to test structure, clarity, and flow. AI can offer alternative wording or styles, but the student remains responsible for selecting language that reflects their own voice and fits the audience.

For hard skills, students should first learn the foundations without relying on AI to not replace their thinking. For example, in coding, students should understand core concepts and logic, then use AI to accelerate application, while auditing the results and learning from what the tool gets right and wrong. Across disciplines, the goal is to use AI to deepen learning, not to outsource it.

Students should not use AI to replace their thinking and substitute their brains, but rather use AI as a strategy to accelerate the learning.

The AI Integrity Check

Before submitting any assignment, students should run a simple integrity review:

a. What did I delegate to AI, and what did I keep manual—and why?

b. Can I explain every step of my answer without looking at my AI tool?

c. Does this work reflect my own voice and understanding, not just polished wording?

d. Can I transfer what I learned to a different situation or problem?

These questions are not designed to restrict AI use. They are designed to ensure that the student's learning is real, defensible, and transferable. These questions should guide students in building confidence in the results while dealing with smarter machines.

3. Assessing with AI

AI also changes how assessment should work. Teachers can use AI to design more personalized assessments based on student needs and gaps. However, assessments must remain equitable, fair, and inclusive.

In the AI-first world, assessment should shift away from measuring whether students can reproduce and retain information toward measuring whether they can reason, apply, and defend conclusions. For example, an exam may take the form of a role-play or case simulation where students use Generative AI to analyze a technical or managerial problem, but the final conclusion requires critical thinking, creativity, and justification of trade-offs.

If assessments do not evolve, the gap will widen between students who use AI effectively and those who do not. A second risk will also increase: students may misuse AI in ways that replace their thinking instead of strengthening it. Clear standards, transparent criteria, and process-based grading can reduce both risks.

4. Employability with AI

AI can also strengthen employability. University career services can use AI to customize coaching and career exploration—helping students map their skills to potential career paths, identify gaps, and build practical plans for improvement. Career centers can also track market shifts in near real time by monitoring emerging skills and job demands.

In the job market of the future, traditional vertical career paths will be less common. Students will need to develop transferable skills—communication, problem solving, leadership, and collaboration—that can move with them across roles. Employers are also placing increasing weight on soft skills, which means students need to become more intentional about self-reflection and learn to communicate their strengths clearly in interviews.

AI can reduce time spent on low-value job-search tasks, such as repeatedly rewriting résumés and filling forms. However, differentiation will come from evidence of capability. Students who can show real projects—experimental work in school, side projects, and applied problem solving—will stand out. Students should also spend more time self-reflecting about their individual skills, and learn how to communicate effectively their experience in order to differentiate themselves in the job interviews process,

Internship placements may decline or stagnate in some industries as entry-level work is automated. That reality increases the importance of demonstration. Students need to be more entrepreneurial in finding projects that set them apart from the crowd and elevate their brand. A student's portfolio, GitHub repository, case write-ups, and AI-augmented projects—combined with strong transferable skills—become a modern credential. We are moving from an era of degrees alone to an era of skills and proof.

The job markets of the future will all be about networking and differentiation. The role of career service will shift from reactive job filling of available opportunities to a proactive and customized experience to each individual student. I would foresee a future where students leverage their experimental projects at school, combined with side projects to apply their knowledge, and these will form the foundation for positioning them in the job markets of the future.

Summary

The Core Shift

- **The old way:** The "sage on the stage" model, where education focused on memorization and a degree served as static proof of knowledge.
- **The new way:** The "coach who guides the tribe" model, where education becomes a laboratory for human–machine collaboration, and value is shown through skills, judgment, and demonstration.

Key Takeaways

- **The new rules of the road:** As AI reduces the effort required for calculation, education must shift toward navigation—ethics, bias awareness, and human–AI coordination.

- **Strategic delegation:** Students should use AI as an accelerator, not a substitute. Learn foundations first, then use AI to apply, test, and extend learning.
- **The AI integrity check:** Before submitting work, students should justify their process through four lenses: delegation choice, step-by-step reasoning, authentic understanding, and transferability.
- **From degrees to skills:** As entry-level tasks are automated, a degree alone is not enough. Differentiation comes from demonstrated ability—projects, portfolios, and transferable skills.

The Reader's Audit

1. In your current study routine, where are you using AI to replace thinking rather than deepen learning? Identify one task you will keep manual this week to strengthen your foundations.

2. If you had to prove your value without listing your degree, what portfolio project, case study, or demonstration could you build that shows real problem-solving ability?

The Next Step

This week: Perform an AI integrity check on your next assignment. Ask yourself, "Can I explain every step of this result without looking at my screen?" If the answer is no, go back to the foundations and rebuild your reasoning. Your value is not the answer AI produced; it is your ability to defend the process and apply it again in a new context.

CHAPTER 9: THE ENTREPRENEURSHIP BUBBLE

Social networks democratized access to opportunities by connecting us with other individuals across the globe who share similar interests. Generative AI is democratizing access to knowledge, which makes me believe that becoming a founder or solo entrepreneur will become more accessible. If large organizations can do more with less, small organizations and individuals will also be able to replicate this model and even move faster. In some non-core support functions, such as marketing and HR, founders can lean on AI agents for the early hiring or marketing strategy before hiring staff that can scale methodology and approaches. Also, building a product is becoming more accessible with vibe coding tools, it takes a few prompts to get an early MVP (Minimum Viable Product) version that used to take weeks or months to build with a team of developers. This new era of

AI-native founders will have a unique opportunity to disrupt markets with little resources. The AI-future will be a place where anyone can start an entrepreneurial adventure, but only a few will succeed in growing that business. Soft skills such as leadership, strategic thinking, and persuasion will differentiate the successful from the unsuccessful ventures. This means that investors need to look more at the human qualities of the founders, such as drive and consistency, rather than the pitch deck and MVP.

When I hear economists talking about the AI bubble, I would rather talk about an entrepreneurship bubble. We used to define a business with a digital identity such as a logo and a brand, a product, and a potential demand in a specific market. Now, anyone will be able to push these boundaries in a matter of hours using Generative AI tools to produce all these elements. But, when can we really create an economic value? AI lowered the cost of building code, generating content, designing products, and analyzing markets; however, the value of entrepreneurship will come from identifying real problems, understanding human pain points, and framing them clearly in order to solve them with AI.

AI has reduced the cost of 'launching a business' to zero. But the cost of 'succeeding' remains as high as ever. Why? Even though AI can build the ship, it cannot generate the wind, which is the passion and the drive that keeps a venture going in challenging environments. Trust will become the new currency of entrepreneurship.

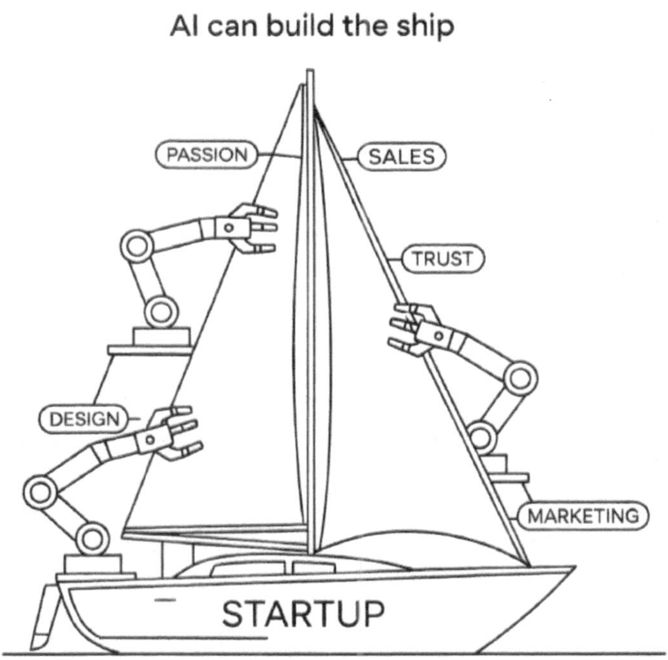

Investors will need to look beyond the revenue, headcount, and inventory to assess trust, reputation, ecosystems, market signals, and client communities. Entrepreneurship is shifting from Technical Founder to Visionary Founder. If any founder can develop a product, then the one who will crack the market is the one who can sell not only to the first customer, but also to the following hundred ones by building a sustainable trust that customers of the future will be prioritizing more than ever. Indeed, there are also deep shifts happening at the customer level; customers are becoming more educated and more aware thanks to AI. Building trust as entrepreneurs will rely heavily on personal brand and reputation in the market. People follow people, and companies who understand this concept early will win market share and grow in a more sustainable way. On the other hand, as jobs are being replaced and automated by Generative AI, and as the lifespan of skills will become shorter, there will be a natural shift towards more entrepreneurial ventures as a way of personal and professional growth. One source of income through employment won't be the norm anymore in the future as most of us will be experimenting some form of instability with shifting jobs and potential job displacements. Nonetheless, before considering entrepreneurship, I would evaluate the following factors that will matter in the future

LOW COST TO START **HIGH COST TO SUCCEED**

1. Tolerance to risk and uncertainty.
2. Understanding of the real problem and pain points.
3. Capacity to keep learning and unlearning.
4. Personal interest and passion for solving the problem.

It's important to pick a problem that we are passionate about, and not only a problem that we think we can solve with AI. The big confusion that everyone is having right now with AI is about understanding the difference between problems that AI can solves (any problem) and problems that we really want to solve and that matters for customers.

That difference will have a direct implication on our consistency and resilience in the projects. This also means that self-awareness will be crucial to help us define what we truly value and where we really want to spend our time, so that we focus our energy on our interests rather than on temporary projects.

Careers in the future will be composed of a mix of corporate and freelance entrepreneurial work, which some could associate with some form of small-scale entrepreneurship, but I believe that each of us should develop this muscle at a young age to build resiliency over periods of economic downturn. If entrepreneurship becomes the new normal across all generations, we should start preparing the youth to embrace change and adaptability, and to develop soft skills that will differentiate them in the entrepreneurial journey. Will everyone become an entrepreneur? Not necessarily, as not all of us can deal with uncertainty and instability, but many of us will need to explore this path as part of their professional lives. This can take multiple forms, such as coaching, a side hustle, and even more. Many studies are showing that the number of founders has increased significantly in the past years, which makes me believe that this trend has already started and will accelerate even more in the AI era with all the shifts coming in the job market.

Summary

The Core Shift

- The Old Way: Starting a business was a high-barrier mission requiring significant capital, technical teams, and months of labor to build a Minimum Viable Product (MVP).
- The New Way: Transitioning to the "AI-Native Founder" who uses "Vibe coding" and agents to build an MVP in hours. The barrier to entry has shifted from technical capability to strategic vision and resilience.

Key Takeaways

- The Entrepreneurship Bubble: AI has reduced the cost of "launching" to nearly zero, creating a flood of generic startups. However, the cost of "succeeding" remains high because AI can build the ship, but it cannot generate the "wind", the human passion and drive.

- From technical to visionary: When any founder can generate code and marketing plans, the differentiator becomes Trust. Personal brand, reputation, and the ability to sell to the "hundredth customer" become more valuable than the initial pitch deck.
- The problem-solution Trap: AI can solve almost any problem, but value only comes from solving the right problems, human pain points that matter to customers. Consistency comes from solving a problem you are actually passionate about.
- The Resilience Muscle: As the one-income employment model fades, entrepreneurship becomes a survival skill. Developing this muscle early allows for agility through shifting jobs and economic downturns.

The Reader's Audit

If you are an aspiring founder or "intrapreneur," conduct this Strategic Review:

The Passion vs. Capacity Test: List three problems you can solve with AI today.

Now, circle the one you actually care about solving. How does your "Life Energy" change when you focus on the circled one?

The risk tolerance audit: On a scale of 1-10, how comfortable are you with uncertainty? If AI provides the "Extension" to lower your financial risk, what is the "Human Skill" (like persuasion or leadership) you need to bridge the gap to success?

The Next Step

This Weekend: Launch a "Micro-Venture." Use a prompt-to-product tool to build a landing page or a basic MVP for a problem you are passionate about. Don't look for perfection; look for a market signal. Does a real human being value your solution? That is the only data point that matters.

CHAPTER 10: AUGMENTED ART

One of the fields raising the most questions about AI is art in its multiple forms (music, movies, book writing, etc.). These fields have already gone through major transformations, moving from physical creation to digital production. That shift made technology part of the creative process—helping artists reach more people, speed up production, and experiment more easily. Generative AI accelerates that pattern by making creation even faster and more accessible.

AI can generate content at scale, but creative value is not defined by volume. When nearly anything can be produced instantly, the work that stands out is defined by intent—clear perspective, strong choices, and a message that resonates with the human soul.

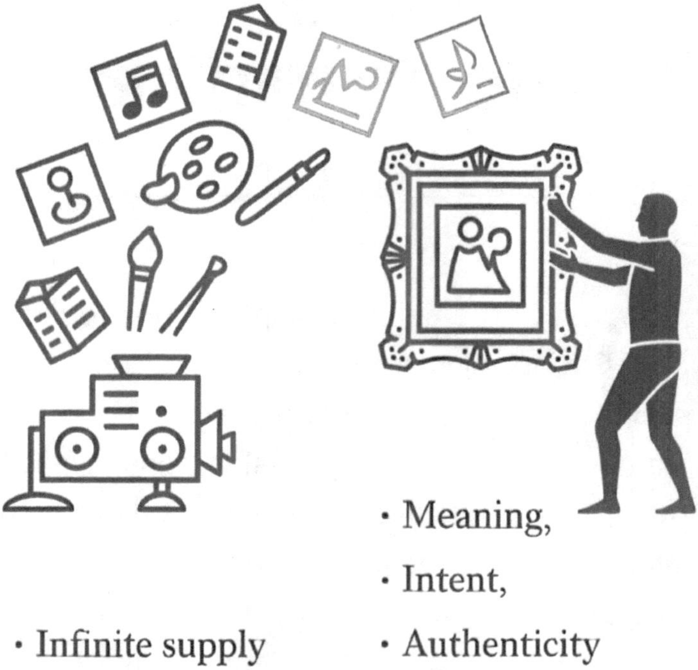

The shift to AI-augmented art shows up clearly in industries like film. AI can reduce the cost of production and shorten the time needed to create drafts of scripts, storyboards, soundtracks, and visual concepts. It also makes it easier for more people who may not have had access to professional tools before to contribute to creative work. In that sense, AI can expand participation and lower the barrier to entry.

At the same time, we should be honest about the question underneath the excitement: if anyone can generate a movie, a soundtrack, or a painting in hours, what creates value? In economics, value is shaped by supply and demand. AI makes supply effectively unlimited, which means scarcity shifts. The scarce element is no longer the ability to generate output; it is the ability to choose what matters, to frame meaning, and to develop a perspective that feels authentic and specific. In an AI-saturated world, the creator who stands out will be the one who can curate, direct, and refine—turning infinite options into one coherent work with a clear point of view.

This is also where copyright and creative ownership matter. AI-generated content is often built from patterns learned from large bodies of existing work. If copyright protections collapse into "anything can be used freely," we risk removing incentives for original creation and undermining trust in creative markets. If we want art to thrive, humans must remain accountable for authorship and responsible for how tools are used. AI can assist, but it should not erase the creator's rights, identity, or ownership of the final work.

There is also a deeper distinction that matters for art: humans create to express meaning—identity, emotion, history, beauty, conflict, and sometimes rebellion. AI generates by recombining patterns from what already exists. It can imitate a style, but it does not live the experiences that give art its weight. Art often comes from the human "glitch"—the imperfect, unpredictable, deeply personal way people interpret life and turn it into something others recognize as true. That is why AI can replicate the brushstrokes of a famous painter, but it cannot fully reproduce the lived context that made the work matter in its time.

Humans create art to express meanings, emotions, identity, traumas, beauty, and rebellion, while AI produce art by combining patterns from existing data. AI can never "rebel" because it is literally built to follow the patterns it was given. Rebellion is a human glitch. That "glitch" is where art lives. AI can synthesize the style of Pablo Picasso, but it can never feel the loneliness that forced him to paint.

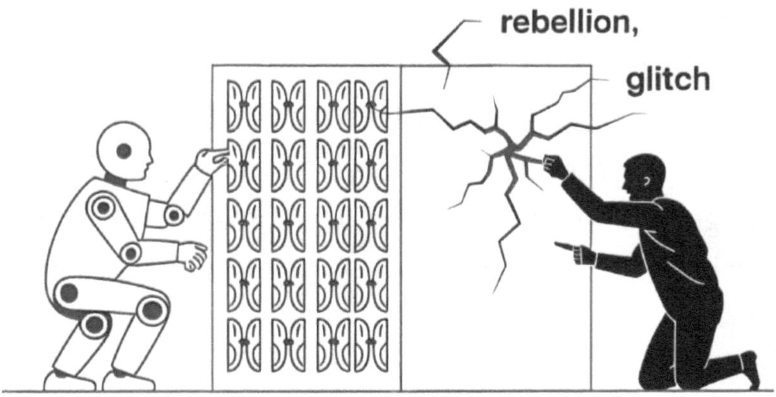

AI follows patterns — humans break them.

Artists of the future, then, will not be those who reject AI. They will be those who use it intentionally—using AI to explore more perspectives, speed production, and test variations—while keeping human direction and meaning at the center. AI can expand the range of options. The artist decides what to say, what to keep, what to cut, and what the work is ultimately trying to communicate.

Music offers a useful picture of how this might feel in practice. AI can lower language barriers, improve accessibility, and expand the ways people experience live performance. Imagine a concert where AI-powered glasses provide real-time translation, where visuals adapt dynamically to the story of a song, or where immersive production elements bring a shared narrative to life. We could revive Michael Jackson with an avatar having the most accurate image and voice representation on stage. Even in those scenarios, the value is not the technology itself. The value is the human experience it supports—the shared attention in the room, the collective memory, and the meaning people attach to the moment.

In short, AI may change how art is produced, distributed, and remixed. But it does not remove what makes art valuable. The core differentiator will remain human intent: the ability to choose a perspective, express meaning, and create an experience that people want to return to—even when infinite content is available.

Would music scene of the future become the playground of artists and fans ? This would one of the most beautiful rewards for AI-augmented art enthusiasts.

Summary

The Core Shift

- **The old way:** Art as a result of physical labor and technical execution, where value was tied to the time spent on the canvas or in the studio.
- **The new way:** Art as a result of intent and orchestration, where value is found in the "human glitch"—the capacity to rebel, to feel, and to move the soul in a world of infinite AI generation.

Key Takeaways

- **The scarcity of intent:** AI can generate everything, but it cannot decide what matters. In a world of infinite supply, the masterpiece is no longer the result of labor; it is the choice of the human who makes sense of the patterns.

- **The rebellion factor:** AI follows patterns; humans break them. Art is often born from rebellion, trauma, beauty, or loneliness—"glitches" that an algorithm cannot feel. AI can synthesize Picasso's style, but it cannot feel the loneliness that shaped it.
- **Copyright as incentive:** For art to thrive, the human must remain the owner. AI is the tool, but the human is the architect of perspective. Protecting the "hand" ensures that creative incentives remain alive.
- **Collective memory:** Technology like holograms or AI-powered visuals are just "extensions." The true value of a concert or a painting in the AI era is the collective memory and the shared human experience it creates.

The Reader's Audit

If you are a creator or an art enthusiast, use this strategic review to audit your creative soul:

- **The "glitch" check:** When you use AI to generate content or art, where is your "human rebellion"? What part of the work is intentionally imperfect or uniquely yours—something an AI could not have predicted?
- **Intent over labor:** If AI removes the manual labor from your creative process (editing, drafting, or mixing), what deeper message or "human why" will you spend that reclaimed time developing?

The Next Step

This week: Become an "orchestrator of perspectives." Use an AI tool to generate five different versions of a creative idea (a poem, a song lyric, or a visual). Do not simply pick the "best" one; combine elements from all of them to create a human masterpiece that reflects your specific identity and memory.

CHAPTER 11: THE MENTAL CAGE

Many people already use AI for deep personal reflection, which can position technology as a companion, therapist, or close digital twin. This use may grow as AI systems are increasingly trained on psychological topics and therapeutic language. However, using AI in this way comes with clear risks. This chapter focuses on three of the most important ones—judgment alteration, confirmation bias, and isolation.

Judgment Alteration, or Losing the "Thinking Muscle"

One major risk is growing dependence on Generative AI in a way that weakens independent thinking and can distort judgment. AI often communicates with confidence—even when it is wrong. Most people have encountered moments when an AI system presented a recommendation as if it were certain, only for that recommendation to be inaccurate or incomplete. If those outputs are accepted without review, decision-making can gradually shift from human reasoning to machine suggestion, and the "thinking muscle" can weaken over time.

Practical safeguard: use AI to generate options, not to make decisions. Treat it as an input, not an authority.

Confirmation Bias, or AI Reinforcing Our Worst Impulses

Decisions that affect relationships require extra care, because AI can unintentionally reinforce a one-sided perspective. A useful guideline is to keep people and values at the center of any relationship decision and evaluate whether the options presented are fair to all stakeholders. AI is strong for transactions (getting facts), but weak for connection (understanding lived context and human nuance).

For example, if you are trying to resolve a dispute with a partner or friend and your prompt is framed in a leading way—such as, "Do you think the other person is responsible?"—the model is likely to generate a response that validates your framing. That may feel helpful in the moment, but it can increase tension and reduce accountability.

A better prompt approach:

"Act as a coach. Based on the situation and my values [X], give an objective perspective on [Y]. List what I may be missing. Propose options that are fair to both sides, and include the risks of each option."

The responsibility remains human: you must evaluate whether the reasoning is balanced, the assumptions are fair, and the outcome aligns with your values.

Isolation, or Replacing Human Warmth with Digital Efficiency

Another risk is a gradual reduction in human interaction. Tools like ChatGPT or Gemini make high-level information and guidance instantly available, which can reduce the need to talk to people when you are processing a challenge. The goal is not to avoid AI, but to keep a healthy balance between machine input and human connection.

AI can be a supplement, but it should not become the main source of support for mental and emotional concerns. A supplement can improve performance, but it cannot replace the "real food" of human relationship, or the expertise of trained practitioners when someone needs real help. No matter how useful AI is for clarifying thoughts, it is important to distinguish between intelligence and presence. AI can generate analysis and language, but it cannot provide the shared human presence that often matters most in moments of pain or uncertainty.

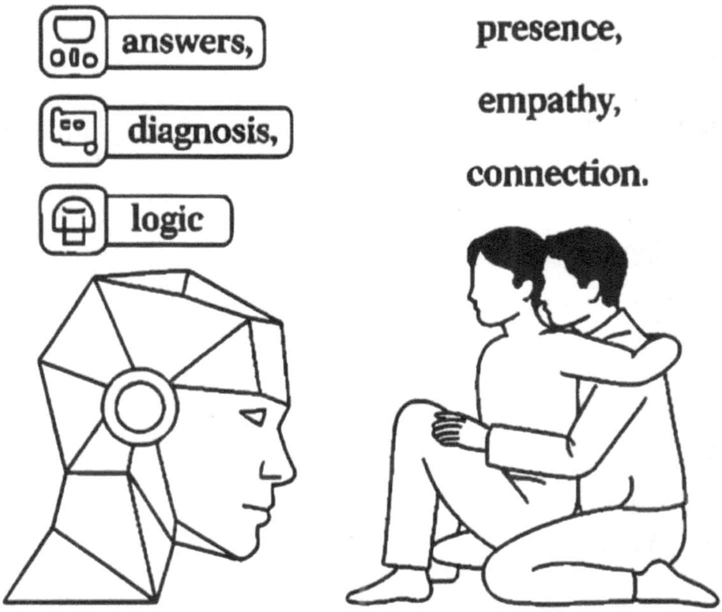

Intelligence is not presence.

A Simple Decision Tool: Pause, Frame, Verify

To use AI without building a mental cage, apply this three-step discipline before you act on an output:

Pause: What decision am I actually trying to make?

Frame: What values, constraints, and stakeholders must be included to make this fair?

Verify: What assumptions, missing facts, or biases could change the answer?

This keeps the tool useful without surrendering your best judgment.

Summary

The Core Shift

- **The old way:** Using AI as a substitute for thinking and treating its confidence as credibility.
- **The new way:** Using AI as a structured input while keeping judgment, fairness, and responsibility human-led.

Key Takeaways

- **Judgment alteration:** If you outsource reasoning too often, you weaken independent decision-making.

- **Confirmation bias:** AI will often reinforce your framing; balanced prompts reduce distorted outcomes.
- **Isolation risk:** AI can clarify thoughts, but it cannot replace human relationship and presence.
- **Pause–frame–verify:** A practical method to keep AI supportive without letting it become directional.

The Reader's Audit

- When you use AI for personal decisions, are you seeking clarity—or seeking validation?
- What is one relationship or situation where you should involve a person before you act, even if AI gives you a clean answer?
- What values must be protected in your decisions (fairness, respect, honesty, long-term trust)?

The Next Step

This week: Before using AI to help with a personal dispute or emotionally loaded decision, write down (1) your values, (2) the strongest argument for the other person's perspective, and (3) what a fair outcome would require. Then ask AI to critique your reasoning—not to validate it.

CHAPTER 12: **ENHANCING COMMUNICATION**

Generative AI is changing how we use language in everyday life. It can help you write more clearly, organize your ideas faster, and communicate with more precision. It can also help you learn new languages more efficiently by reducing the friction of practice—giving you consistent repetition, real-time feedback, and realistic conversation scenarios on demand.

This chapter is about both. First, we will look at how to use AI to strengthen your writing without losing your voice. Then we will focus on using AI to build language skills faster through practice that is more consistent, more personalized, and easier to sustain than traditional methods alone.

Used well, AI can be an amazing language partner, helping you draft, revise, practice, and improve your communication.

1. Writing

Generative AI has been trained heavily on English-language content and was adopted earliest at scale by English-speaking markets. That does not mean it has been trained "wrongly," but it does raise a reasonable concern: the default style, references, and tone of many AI outputs can lean toward being Western and standardized. Over time, that can produce cultural flattening—communication that sounds polished but increasingly similar across people, industries, and countries.

I notice this most when I use AI to draft emails. The language often comes back "perfect"—clean, direct, and professional. But that perfection raises an important question: if most communication becomes optimized and uniform, what happens to cultural voice, personal nuance, and human connection?

This is where responsibility returns to the user. If we accept AI outputs without revision, we risk trading individuality for convenience. Cultural diversity plays a real role in how societies innovate, collaborate, and build meaning, so we must strive to maintain our culture. The goal is not to reject AI, but to use it without surrendering what makes communication human. A simple rule helps: if your message is perfectly written but it does not sound like you, you have not communicated—you have only transmitted information.

The 90/10 Rule for Keeping Your Voice

One practical approach is to use AI for structure and efficiency while keeping a meaningful portion of the message human. In email drafting, for example, you can let AI produce 90 percent of the framework—clarity, organization, and completeness—while you deliberately keep 10 percent for your own phrasing, examples, cultural references, tone, or metaphors. That human signal is what keeps the message personal and relational rather than generic.

This approach improves productivity while keeping you mentally engaged and accountable for what you send.

2. Language Learning

There is a second side to this story, and it is more optimistic. AI is democratizing access to language learning. Instead of using AI to flatten culture into one dominant communication style, we can use it to immerse ourselves in other cultures through language—more quickly, more affordably, and more consistently than traditional methods allow.

Since my early use of AI, I have found language learning to be one of the most valuable and practical use cases—especially when it complements in-person classes. To show what this looks like beyond theory, the following case study illustrates how AI can reduce the friction of practice, increase feedback frequency, and help sustain momentum over time.

Case Study: From Four to Seven Languages

The Challenge: Mastering new languages while working full time typically takes years of study—often with inconsistent sessions, heavy repetition, and feedback that arrives too late to correct mistakes in real time.

The Strategy: Move from passive learning (textbooks and occasional practice) to AI-supported immersion with real-time feedback and personalized content.

- **Vocabulary:** I use AI-powered language applications such as LingQ to import articles from the web, analyze the content, and save new words for review. I choose content I genuinely care about—technology and business articles, as well as lyrics from music I enjoy—because relevance keeps engagement high. AI's value here is personalization: it adapts the learning journey to your interests and pace.
- **Speaking:** I use tools such as TalkPal or ChatGPT to practice real-life scenarios—introductions with colleagues, shopping,

ordering at restaurants, and other everyday interactions. As I practice, the tool provides immediate feedback on vocabulary, pronunciation, grammar, and sentence structure. This makes language tutoring more accessible and more frequent than traditional models.

- **Writing:** I use tools such as TalkPal, ChatGPT, or Gemini to practice writing in informal conversation and formal essays. In both cases, AI provides fast, detailed feedback and helps me correct patterns early. The result is faster progress and clearer awareness of gaps.

Comparing Traditional Language Learning vs. AI-Augmented Learning

Content discovery

- **Traditional:** Relies on generic textbooks and standardized lessons that may not match your interests.

- **AI-augmented:** Uses interest-driven content (articles, videos, lyrics, and work-related material) tailored to your level and goals.

Practice

- **Traditional:** Limited practice time, often tied to class schedules or a weekly tutor session.
- **AI-augmented:** On-demand practice, including real-life scenarios and role-play, available any time you have a few minutes.

Feedback speed

- **Traditional:** Corrections arrive later—after class, after homework review, or at the next session.
- **AI-augmented:** Real-time feedback on vocabulary, grammar, structure, and pronunciation while you practice.

Fluency timeline

- **Traditional:** Progress is steady but slow, often two to three years per language for working adults.
- **AI-augmented:** Faster iteration and more repetition can reduce the timeline to months, depending on consistency and intensity.

In the past, improving communication usually meant slow cycles—drafting, second-guessing, waiting for feedback, and practicing only when you had time or a live partner. With AI, communication becomes a daily lab: you can shape clearer messages, test tone and structure, practice conversations, and get immediate feedback while still keeping your voice and intent in control. Used well, AI does not replace human expression; it strengthens it—helping you communicate with more clarity, consistency, and confidence across both writing and language learning.

Summary

The Core Shift

- **The old way:** Communication and language improvement depended on slow cycles—write, wait, revise—often with limited feedback and inconsistent practice.
- **The new way:** AI reduces the friction of improvement. You can draft faster, get feedback immediately, and practice communication daily—while still keeping human intent and voice in control.

Key Takeaways

- **AI as a practice partner:** Used well, AI turns communication into a daily skill, not an occasional project—drafting, refining, and practicing in short sessions.
- **The 90/10 rule:** Let AI handle structure and clarity, but keep a meaningful portion for your own voice (your phrasing, examples, tone, and cultural nuance) so the message sounds like you.

- **Fast feedback changes everything:** Language grows faster when feedback is frequent. AI makes correction and iteration immediate, which helps you build momentum.
- **Immersion without the barriers:** AI-supported role-play and scenario practice make "real-life" communication available on demand, which strengthens confidence and fluency.

The Reader's Audit

1. **The Voice Check:** Think about your most recent AI-assisted message (email, post, or response). Did it sound like you? What one adjustment would make it more natural—tone, phrasing, or a specific example?
2. **The Practice Check:** Which area would improve your communication fastest right now: writing, speaking, or vocabulary? What would "fifteen minutes a day" of practice look like in that area?

The Next Step

This week, run a simple communication routine: use AI to draft one message (email, summary, or note), then revise it using the 90/10 Rule so your intent and voice are clear. Next, practice one short language scenario (a brief role-play, an introduction, or a real-life situation) and ask for feedback you can apply immediately.

CHAPTER 13: PRACTICAL APPLICATIONS

If you have made it to this chapter, you have already covered the foundations of AI literacy that matter most over the next five to ten years: how AI is changing work and learning, what it can support, and what must remain human-led. Now we shift from theory to execution. In this chapter, I share practical, repeatable use cases from my own life—across finance, career, branding, learning, and health—that you can adapt to your situation. Whether you choose to use strong prompting or lightweight AI agents, when using Generative AI, the goal is the same: to reduce friction, move faster, and make better decisions—while keeping judgment, accountability, and values in human hands.

We will explore five application areas:

1. Finance
2. Career
3. Branding
4. Learning
5. Health

AI as an amplifier across life domains

1. Finance

I use AI to clarify goals, explore scenarios, and improve the quality of my research. I treat it as a research accelerator, not a decision-maker. It can summarize, compare, and surface questions I may have missed, but it cannot decide what is right for my goals, risk tolerance, and time horizon.

Here is how it looks in practice:

A Practical Workflow:

Clarify → Compare → Pressure-Test

Step 1: Clarify the decision. What am I actually deciding?

Step 2: Compare options. What are the realistic paths?

Step 3: Stress-test assumptions. What could make this wrong?

Sample Prompt: Financial Decision Support

"Act as a financial research assistant. I am considering [decision]. My context: [age], [goal], [time horizon], [risk tolerance], [income stability], [current assets], [constraints].

1. List the key questions I should answer before deciding.
2. Provide three scenarios (conservative, base, aggressive) with assumptions clearly stated.
3. Identify the biggest risks and what would change your recommendation.
4. Give me a checklist of what to verify using primary sources."

Real Estate: Use AI for Preparation, Not Proof

I use AI similarly in real estate negotiations. It can help estimate a reasonable range using publicly available information, draft offer language, and compare scenarios using assumptions I provide. The practical value is preparation: it helps me validate basic math, ask better questions, and spot claims that do not hold up logically.

Sample Prompt: Offer and Negotiation Prep

"Act as my real estate negotiation prep assistant. Property: [location], [price], [type], [key facts]. My goals: [primary residence/investment], [budget], [timeline], [non-negotiables].

1. List the key due diligence questions I should ask.
2. Identify assumptions that could make the deal look better than it is.
3. Draft an offer email with clear terms and respectful tone.
4. Draft two counteroffer options: one firm, one flexible."

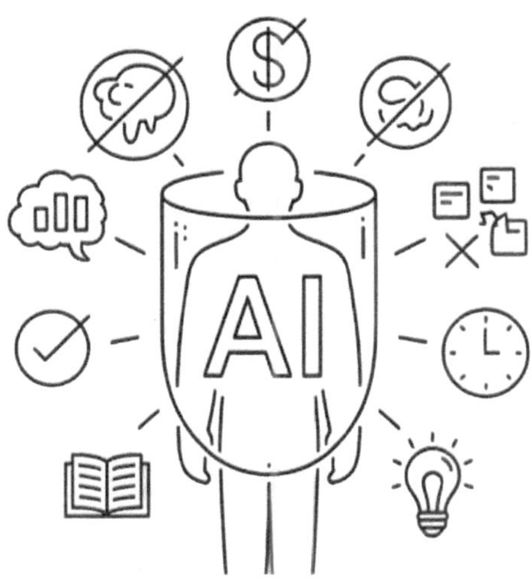

Ai as an intellectual shield,
not a decision-maker

2. Career

Career use cases range from creating résumé drafts that are ATS-friendly to preparing for interviews through role-play. AI can also support leadership development by helping you draft communication, rehearse difficult conversations, and structure decisions.

One of the most practical shifts is this: do not only apply for a job—simulate it.

A Practical Workflow:

Tailor → Rehearse → Refine

Step 1: Tailor your résumé to one role.

Step 2: Rehearse the role through scenarios.

Step 3: Refine your stories until they sound like you.

Sample Prompt: ATS-Friendly Résumé (Without Losing Voice)

"Here is my résumé: [paste]. Here is the job description: [paste].

1. Identify the top eight keywords and responsibilities I must match.
2. Rewrite my résumé bullets to align, but do not exaggerate or invent.
3. Keep my tone professional and plainspoken.
4. Flag any gaps and suggest what I should address in the interview."

Sample Prompt: Role-Play Interview Simulation

"Act as a hiring manager for this role. Ask me ten questions, including three scenario questions and two conflict questions. After each answer, critique my clarity, specificity, and impact. Push for details when I get vague."

3. Branding

AI can help with content planning, audience segmentation ideas, drafting, editing, and repurposing content across platforms. It can also help outline a basic website or generate early structure for brand materials.

The boundary is simple: AI can help you package and scale your ideas, but it cannot define them for you.

A Practical Workflow:

Point of View → Content → Repurpose

Step 1: Define what you stand for.

Step 2: Draft content with AI.

Step 3: Rewrite the final version so it sounds like you.

Sample Prompt: Preserve Voice While Drafting

"My topic is [topic]. My audience is [audience]. My point of view is [belief]. My tone is [tone]. Draft three versions:

1. Short social post
2. Longer LinkedIn-style post
3. Email to my audience.
   Then give me a "voice checklist" so I can rewrite the final version in my own words.

A practical rule: let AI draft structure and options, and then you decide what you stand behind. AI can be the megaphone, but you remain the voice.

4. Learning

AI can reduce the friction of learning when you treat it as a tutor and practice partner rather than an answer engine. Used well, it turns learning into a loop: you attempt, you test, you refine, and you repeat.

A Practical Workflow:

Explain → Practice → Quiz

Step 1: Ask for a simple explanation.

Step 2: Do practice exercises.

Step 3: Get quizzed until it sticks.

Sample Prompt: Skill-Building Loop

"Teach me [skill] as if I'm a beginner.

1. Explain the concept simply.
2. Give me five practice tasks that increase in difficulty.
3. After each task, grade my answer and explain the correction.
4. Quiz me tomorrow with ten questions based on what I struggled with."

For nontechnical skills, AI can also help you rehearse: public speaking, leadership communication, tough conversations, and language practice.

The rule stays the same: AI can accelerate practice, but you still need to perform the skill without the tool when it counts.

5. Health

I use AI to organize information and improve self-awareness. It can help summarize lab results, track patterns over time, and generate questions to ask a provider. The value is clarity, especially when health information feels overwhelming.

However, keep in mind: AI does not examine patients, does not carry legal responsibility, and does not know your full medical context. I use it to prepare for better conversations with clinicians, not to replace them.

A Practical Workflow:

Organize → Question → Confirm

Step 1: Organize what you know.

Step 2: Generate questions for your provider.

Step 3: Confirm decisions with qualified professionals.

Sample Prompt: Appointment Prep

"Help me prepare for a medical appointment. Here is what I know: [symptoms], [timeline], [labs/notes], [medications], [history].

1. Summarize this into a clear one-page briefing I can bring to my doctor.
2. List ten questions to ask, prioritized by importance.
3. Identify what information is missing that would change the assessment.
4. Explain what is urgent vs. what can wait."

In practice, the value of Generative AI is not that it replaces your work—it is that it reduces friction in the parts of life and work that drain time and attention. Used well, it becomes a support system for clearer thinking, faster preparation, and more consistent execution across finance, career development, personal branding, learning, and health. The boundary is the same in every area: AI can accelerate research, drafting, and organization, but judgment, accountability, and values must stay human-led. That is how you get the speed without losing the responsibility.

Summary

The Core Shift

- **The old way:** Treating AI like an answer machine—using it for quick outputs without clear context, without verification, and without owning the consequences.
- **The new way:** Treating AI like a practical assistant—using it to reduce friction and raise decision quality while keeping judgment, accountability, and values in human hands.

Key Takeaways

- **Use AI for preparation, not permission.** Let it help you research, compare options, and draft materials—but you decide what is right and what you will stand behind.
- **Context is the difference between noise and value.** The better your inputs (goals, constraints, audience, tone, and assumptions), the more useful the outputs.

- **Require assumptions and options, not a single "best" answer.** In finance, career, and health especially, ask for scenarios, tradeoffs, and risks.
- **Keep a human signature.** In branding and communication, AI can polish structure, but your point of view, voice, and values must remain yours.
- **AI accelerates learning only when you still do the work.** Use it to practice, get feedback, and iterate—but protect the thinking that builds real competence.

The Reader's Audit

Answer these in the book.

1. Where do I need speed most right now: research, drafting, organizing, or practice?
2. Which area would improve fastest if I reduced friction by 20 percent: finance, career, branding, learning, or health? Why?

3. What is my non-negotiable boundary with AI in this area? (Examples: "I verify sources," "I never send without revision," "I never treat output as medical advice," "I always compare scenarios.")

4. What decision do I keep avoiding because it feels overwhelming, and how could AI help me prepare for it without making it for me?

The Next Step

This week, pick **one** of the five areas and run a simple "assist, then decide" cycle:

1. **Assist:** Use AI to Generate a structured starting point (questions, options, and a draft plan).

2. **Verify:** Check assumptions, identify what is missing, and confirm anything high-stakes with reliable sources or professionals.

3. **Decide:** Make the final call yourself—and document the reason you chose that path.

If you want a concrete starting point, choose the one that fits your life right now:

- **Finance:** Ask for three scenarios based on your risk tolerance and time horizon, then validate the assumptions before acting.
- **Career:** Run a mock interview and refine your stories until they sound like you and match the role.
- **Branding:** Draft one piece of content with AI, then rewrite the final 10–20 percent in your voice and publish it.
- **Learning:** Use AI to generate a practice routine and feedback loop, then do three short practice sessions this week.
- **Health:** Summarize your information and write five questions for your next appointment—but keep diagnosis and decisions with your clinician.

CHAPTER 14: CHANGE MANAGEMENT

It is difficult to imagine an AI future without change management, because one of the most important skills of the future will be our ability to adapt to continuous change. As a certified change practitioner, I want to introduce a few core concepts and give you a clear place to start.

First, people change when the cost of staying the same becomes higher than the cost of change. We do not always change when we see the light; we change when we feel the heat. In the context of AI, that "heat" often shows up as concern about the longevity of our jobs, the expiration of our skills, or the fear of becoming irrelevant.

To make change more practical, I use a simple framework: **USPR (Understand, Sell, Practice, Repeat).**

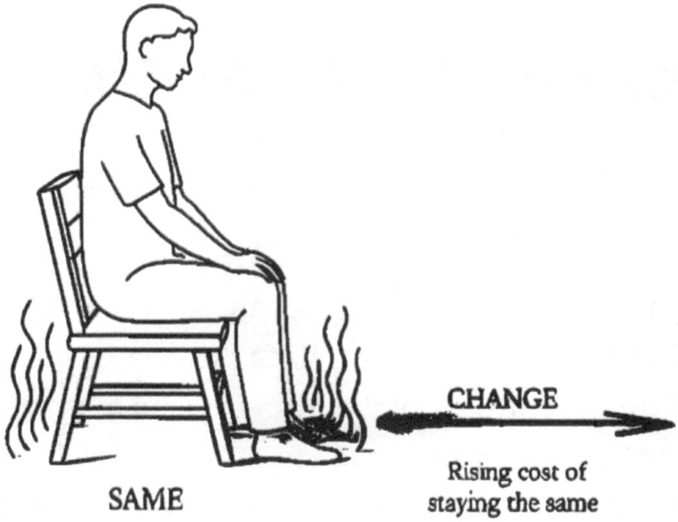

We don't change when we see the light.
We change when we feel the heat.

Breakdown of the USPR Framework

U (Understand)

The first step is to understand what adjustments you need to make to navigate this new environment. That starts with clarity: why AI matters in your world, what is actually changing in your role, and what is not. Pay attention to the signals your organization sends.

Leaders should communicate direction, repeat the message enough for it to stick, and be honest about risks. If those things are not clearly provided, you should seek them out so fear doesn't preside over good judgement.

S (Sell to Yourself)

Some shifts do not require motivation speeches. They require a clear, personal reason to move. With AI, that reason often becomes practical before it becomes philosophical. The pressure shows up in small ways: work that takes longer than it should, tasks that drain energy, and a growing sense that the environment is changing faster than your habits. The point is not to adopt AI because it is trending. The point is to understand what staying the same will cost you—and what small changes could give back.

People also respond to change differently, and that is normal. Some individuals will experiment early, replacing parts of their workflow and quickly seeing gains in speed and quality. Others will hesitate, not because they lack capability, but because uncertainty creates friction.

Over time, resistance often softens when the benefits become concrete and observable.

The real requirement is clarity. You need a personal answer to a simple question: What is in it for me? For some, the answer is time saved and fewer repetitive tasks. For others, the first response is concern—especially the fear of becoming less relevant or being replaced. When that fear appears, the best response is not avoidance. It is a small, controlled experiment. Observe one practical use case that produces real value for someone else, then test one version of it in your own work. Start with a low-risk task, compare the results to your usual approach, and decide based on evidence.

This is how change becomes manageable. You do not need to adopt everything. You need to identify where AI genuinely improves your outcomes—and where your judgment must remain in control.

P (Practice)

Once awareness and desire are present, the next step is competence. You build capability by learning how to use Generative AI, work with agents, and keep learning as the technology evolves.

For some people, that means formal education. For others, it means online learning and practical experimentation. The point is to find your rhythm. If you do not know how to work in the new way, adoption stays fragile.

R (Repeat)

Practice becomes progress when it becomes a habit. Over time, your proficiency grows, you feel the benefits, and you can even teach others. Then the cycle starts again, because the environment keeps changing.

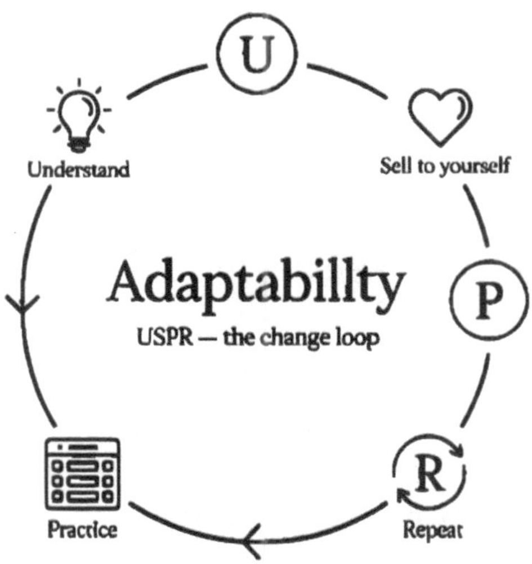

In the AI era, the most stable thing you can be is adaptable.

Summary

The Core Shift

- **The old way:** Viewing change as a one-time event—a project with a start date and an end date.
- **The new way:** Treating change as a continuous cycle, where adaptability is a long-term advantage.

Key Takeaways

- **The heat of change:** People do not change because they see the light; they change when they feel the heat. When the cost of staying the same exceeds the cost of transition, change becomes possible.
- **The USPR loop:** A practical approach to AI adoption is iterative: Understand the shift, sell it to yourself, practice the new way, and repeat until it becomes normal.
- **Repetition builds safety:** Change triggers fear. Clear communication, repeated consistently, helps people move from threat to clarity.

- **Early adopters create proof:** If you feel stuck, borrow momentum. Watch what is working for early adopters and replicate one specific use case.

The Reader's Audit

- **The mirror test:** What is the biggest "heat" you feel about AI right now—job longevity, skill expiration, competition, or something else? What would make the change feel like a gain instead of a loss?
- **The resistance check:** Name one AI tool or concept you have been resisting. How many real attempts have you made to understand it? Is your resistance based on experience and data, or on discomfort and uncertainty?

The Next Step

This week, move into the **practice** phase. Identify one early adopter in your network or industry. Ask them for one small, specific "win" they have had with AI. Replicate that exact use case three times this week. Do not aim for mastery—aim for repetition. Repetition is how resistance turns into capability.

CLOSING

If you have journeyed through these fourteen chapters, you are no longer just a spectator of the AI era—you are shaping how it shows up in your life and work.

We began by defining AI as an extension: a tool that can amplify speed, scale, and analysis. Along the way, we explored how to use it as an intellectual shield against noise, how to reduce the energy leaks that drain purpose, and how to reclaim time that was once consumed by the "Age of Keyboards." As you close this manual, I want to leave you with the most important truth of all: AI is the pen, but you are the author—the hand that controls it.

The Final Boundary

We are the first generation to live alongside an intelligence that is not our own. In the presence of a system that never sleeps and never forgets, it is easy to feel replaceable.

The boundary is this:

- AI can generate output, but it cannot own intent, values, or responsibility.
- AI can produce answers, but you decide the question worth asking.
- AI can find patterns, but you decide what matters in context.
- AI can offer options, but you choose—and live with the consequences.
- AI can scale execution, but you protect trust, judgment, and direction.

The machine can synthesize the style of a masterpiece, but the meaning of the work still comes from the human who chooses the message, the perspective, and the risk.

Your New Mission

As you step back into your life, do not fear the "Entrepreneurship Bubble" or the "Purpose Crisis." Treat them as signals that a new kind of leadership is required—one that is grounded in clarity, character, and contribution.

The goal of this era is to use machines for what they do well, while strengthening what only humans can carry: judgment, accountability, relationships, and purpose. Every day, you will face a choice: default to digital efficiency, or use the extension with intention.

The goal of this new era is not to become more like a machine, perfect, synthetic, and cold. The goal is to become more human than ever.

To be vulnerable, to be culturally unique, and to be fiercely authentic.

Reclaim your time. Protect your voice. Charge your battery. Stay imperfectly human.

The future is built one decision at a time—one prompt at a time. The extension is ready.

Now, show the world your hand.

AUTHOR BIOGRAPHY

Mohammed Benzakour is a seasoned Digital Strategy Leader with 10+ years of experience across Europe and the Middle East. Operating at the intersection of technology and human potential, he has held senior roles at global organizations including LinkedIn, Microsoft, and Deloitte. Throughout his career, he has served as a trusted advisor to governments and corporate leaders as they navigate workforce transformation and future-ready strategies, and is a frequent speaker at international conferences on AI, leadership, and the future of work.

As a Certified Change Practitioner and an active member of the International Coaching Federation (ICF), Mohammed focuses on the psychological and cultural shifts required to thrive in a machine-driven world. He is a true digital polymath who leveraged AI to expand his linguistic repertoire from four to seven languages, proving that technology is a powerful extension of human capability, not a replacement for it.

Ranked in the top 0.01 percent of AI users globally, Mohammed has built more than one hundred AI agents to stress-test the boundaries of productivity and intent. He remains dedicated to empowering professionals to reclaim their time and protect their authentic voice in the Age of Intent.

www.ingramcontent.com/pod-product-compliance
Lightning Source LLC
LaVergne TN
LVHW090950250926
848814LV00061B/2

* 9 7 9 1 0 9 8 2 8 7 7 2 5 *